Graham Walker is a Professor of Mechanical Engineering at the University of Calgary, Alberta, Canada. He gained his B.Sc. and Ph.D. at King's College, Newcastle upon Tyne, University of Durham. He has also worked at the Illinois Institute of Technology, Chicago, and Imperial College of Science and Technology, University of London. Professor Walker served an engineering apprenticeship in British Railways Workshops, had military experience with REME in the Egyptian Canal Zone, and industrial experience with the De Havilland Engine Company and British Timken. He has served as a consultant on Stirling engines to government departments and industrial companies in Great Britain, the United States of America, and Canada.

Stirling-cycle machines

2-kW Electric Generating set incorporating a Philips 1–98 Stirling-cycle engine. (Courtesy United Stirling A.B., Malmo, Sweden.)

Stirling-cycle machines

G. WALKER
University of Calgary, Canada

CLARENDON PRESS · OXFORD
1973

Oxford University Press, Ely House, London W. I
GLASGOW NEW YORK TORONTO MELBOURNE WELLINGTON
CAPE TOWN IBADAN NAIROBI DAR ES SALAAM LUSAKA ADDIS ABABA
DELHI BOMBAY CALCUTTA MADRAS KARACHI LAHORE DACCA
KUALA LUMPUR SINGAPORE HONG KONG TOKYO

Printed in Great Britain by The Pitman Press, Bath

Preface

THE last decade has witnessed a resumption of interest in Stirling-Cycle machines. They are widely used in a variety of types and sizes for small-scale cryogenic cooling purposes. Prototype engines of advanced design are being evaluated for automobile application.

The renaissance of the Stirling engine, both as a cooling machine and prime mover, is almost entirely due to the accomplishments of workers at the Philips Research Laboratories, Eindhoven, Holland, in an intensive research and development programme extending back to the 1930s. As a result of this effort Philips and their licensees are so far ahead in the technology of Stirling engines as to appear virtually unassailable on the commercial front. Possibly this apparent domination of the field has tended to inhibit other major independent development efforts, yet there appear to be alternatives to the Philips family of machines that have the potential for application in fields which in no way conflict with those chosen for development by Philips.

One difficulty confronting newcomers to the field is the lack of any kind of reference work. Much information is available, but it is widely scattered over many journals, and its collection and classification is a time-consuming task. This book is my attempt to help newcomers see not only the wood but also the trees, it represents a distillation of my experience and interest in Stirling engines, since 1956. It is by no means a comprehensive survey of the subject, but will I hope prove useful both to students and to practising mechanical engineers.

I have to acknowledge much help. I am grateful to Aubrey F. Burstall, Professor Emeritus of the University of Newcastle upon Tyne, who introduced me to my first Stirling engine. Theodor Finkelstein, Rolf Meijer, and J. W. L. Kohler have all contributed to my understanding and appreciation of the subject. More recently, William Beale, Jack Roberts, John Kentfield, Horace Rainbow, and Ken Round have helped me. I am grateful to Stig Carlqvist for the opportunity to see the work in progress at United Stirling. By his interest and enthusiasm Frank Wallace, Head of Engineering at the University of Bath, helped greatly in the production of this book, which is based on my notes for a seminar on Stirling engines, given during my stay at Bath in 1971–2. Finally, I am grateful to my wife whose unfailing good humour has made the writing of this book possible.

G. W.

Bath. June, 1972.

Contents

1 Introduction

Definition

A Stirling-cycle machine is a device which operates on a *closed* regenerative thermodynamic cycle, with cyclic compression and expansion of the working fluid at different temperature levels, and where the flow is controlled by volume changes, so that there is a net conversion of heat to work or vice versa.

Machines exist which operate on an *open* regenerative cycle, where the flow of working fluid is controlled by valves. For convenience, these may be called Ericsson-cycle machines, but, in practice, the distinction is not widely established, and the name Stirling engine is frequently indiscriminately applied to all types of regenerative machine. The generalized definition embraces a large family of machines with different functions, characteristics, and configurations. It includes both rotary and reciprocating machines, utilizing mechanisms of varying complexity. It covers machines capable of operating as prime movers, heat pumps, refrigerating engines, or pressure generators.

Nomenclature

Stirling engines are frequently called by other names, including hot-air or hot-gas engines, or one of a number of designations reserved for particular arrangements of engine, i.e. Heinrici, Robinson, or Rankine–Napier. The result is a general lack of clarity in the nomenclature. It may be argued, convincingly, that the designation 'Stirling cycle' should be reserved for a particular idealized thermodynamic cycle, and the name 'Stirling engine' for a particular form of machine (which, incidentally, *does not* work on the Stirling cycle, a situation that does nothing to improve clarity). A preferred generic title would be 'regenerative thermal machine', but it is almost certainly too late for logic to prevail, and the name 'Stirling engine' will continue to be used widely and indiscriminately. A clear distinction should always be made between machines where the flow is controlled by (a) volume changes (Stirling engines) and (b) valves (Ericsson engines), because they have radically different characteristics.

Early history

Stirling and Ericsson engines have a long history, which has been well surveyed by Finkelstein (1959).†

† See Bibliography for complete reference.

Some late eighteenth century machines can be recognized as embryonic hot-air engines, but the principal developments occurred in the early nineteenth century. The first engine to work properly was probably the open-cycle hot-air engine built by Sir George Cayley in 1807. Robert Stirling, a Minister of the Church of Scotland, invented the closed-cycle regenerative engine in about 1816. Later, the Swedish inventor John Ericsson, working in England, introduced the open-cycle regenerative engine. Subsequently, in Britain, Europe, and the U.S.A., thousands of engines, in a variety of shapes and sizes, were widely used, throughout the whole of the nineteenth century. They were reliable, reasonably efficient and, most important, safe, compared with contemporary reciprocating steam-engine installations. Although most machines were small (of quarter to five horsepower), some large machines were made, also. Perhaps the most notable was the marine engine, built by Ericsson in 1853, having four cylinders which were 14 ft in diameter, with a stroke of 5 ft, running at 9 rev/min, and producing about 300 b.h.p. This was installed in a ship called 'The Ericsson', which capsized in a squall in New York harbour.

About the middle of the nineteenth century, the invention of the internal-combustion engine, in the form of the gas engine, and its subsequent development as a gasoline- and oil-fuelled engine, along with the invention of the electric motor, caused the use of Stirling engines to largely diminish until, by 1914, they were no longer available commercially in any quantity. However, production of machines for special purposes (e.g. kerosene-burning machines operating fans, for use in tropical countries) lingered on in England until at least 1946, and model engines are still being made.

Philips engine

Research on Stirling engines commenced at the Philips Laboratories, Eindhoven, in the late 1930s, and has been in progress continuously since that time. Initially, the work was directed to the development of small thermal-power electric generators for radios and similar equipment, for use in remote areas, where storage batteries were not readily available. Subsequent development of radio valves and batteries and, particularly, the introduction of transistors relaxed the requirement for small electric-power generators. However, sufficient encouraging progress had been made for research to continue, with emphasis moving to engines of higher power. Studies have embraced the experimental development of engines of various sizes up to 450 h.p., with attractive characteristics, compared with conventional internal-combustion engines. Principal among these is the low noise and air-pollution level of the Philips engine, at a thermal efficiency and specific output comparable, or better than, a petrol or oil engine. This combination, assisted by increased public concern about the environment, has concentrated attention on the use of the Stirling engine for automotive application. Intensive research and development is in progress on advanced prototype vehicle engines and their associated systems.

The course of the Philips engine development has been described by Meijer (1969*a*).

Other work on Stirling engines was carried out by General Motors, operating as a licensee of Philips from 1958 to 1970, as detailed by Heffner (1965). More recent licensees of Philips include the 'Entwicklungsgruppe Stirling Motor M.A.N. – M.W.M.', formed in West Germany in 1967, and the Swedish consortium 'United Stirling AB', established in 1968. Brief details of the interests of these two companies were given by Neelen *et al.* (1971). Other licence agreements are thought to be under discussion.

Refrigerating machines

Stirling engines operate well as cooling engines. The possibilities of this were recognized, as early as 1834, by John Herschel, and, in 1876, Alexander Kirk described a machine that had been in use for ten years. However, it was not until the late 1940s that serious effort was directed to the commercial development of Stirling-cycle cooling engines. Again, this was undertaken by the Philips Company, at Eindhoven. The first cooling engine (an air liquifier) was introduced in 1955. Since that time, further research has resulted in the development of a variety of cryogenic cooling engines, covering a wide range of cooling capacities, and has led to the manufacture of associated equipment for cryogenic research and industrial applications. So far, Stirling-cycle cooling engines have proved more suitable for the cryogenic (extremely low temperature) range, rather than the higher temperature-range (of domestic and industrial interest) which is dominated, at present, by 'Freon' vapour-compression refrigerating machines.

Other manufacturers have entered the small (and miniature) cryogenic cooling engine market, including Malakar Labs. Inc. and Hughes Aircraft Co., in the United States. These companies, together with North American Philips Inc. (who specialize in miniature cryogenic coolers), have as their principal interest the provision of small cooling engines for electronic applications, mainly in infra-red detection equipment for a variety of military and civil purposes.

Other reciprocating regenerative cryogenic cooling engines have been developed, principally the Collins helium liquefier, by A. D. Little Inc., and a variety of Gifford–McMahon machines. All these machines have valves and, in accordance with the definition adopted earlier, must be classified as Ericsson-cycle machines, and are, thus, beyond our field of interest. This is not to suggest that such machines are unimportant. The development, by Samuel Collins at M.I.T., of a simple inexpensive reliable expansion-engine, capable of liquefying helium, was among the most significant advances in cryogenic engineering, opening up the possibilities of helium research on a broad front. The future benefits of this broadening of research in terms of superconducting electric-power transmission and miniaturized electronics are incalculable.

Many other studies of Stirling engines have been made, but have not led to any commercial applications, although significant contributions, principally academic, have been made to the literature of Stirling engines by Finkelstein, Smith, and Walker.

2 Ideal thermodynamic cycles

Some elementary considerations

The first and second Laws of Thermodynamics appear to apply to all thermal-power machines, including Stirling engines.

The first Law of Thermodynamics

The first Law, which is a restatement of the Law of Conservation of Energy, denies the possibility of an engine (or some thermodynamic 'black-box') to exist, from which power, or work, can be drawn continuously, without replenishment. The first Law requires that at least as much energy (in any form) shall be *supplied* to the machine as is *taken* from it. Let us consider air and petrol, supplied to a spark-ignition engine. Firstly, they combine in a combustion process, and the hot gases drive the engine. Of the energy supplied in the fuel, about one-third goes to useful work output from the engine, another third goes to the cooling system, and the remaining third leaves the exhaust as low-grade thermal energy. If the petrol supply is terminated, the engine stops. This is a direct application of the first Law of Thermodynamics, and a matter of common experience.

The second Law of Thermodynamics

The second Law of Thermodynamics is, perhaps, less well understood. One statement of this Law is that it is not possible to construct a system which will operate in a cycle, extract heat from a reservoir and do an equivalent amount of work on the surroundings. The first Law says that the work produced can *never be greater* than the supplied heat, while the second Law goes further, and says that it *must always be less.* In the spark-ignition engine, it is the second Law which denies the possibility of converting all the energy in the supplied petrol to useful work. Some of the energy must be 'wasted' in the form of heat which is rejected to the cooling system or the exhaust.

These bold statements will suffice for our purpose here. For fuller discussion of the first and second Laws of Thermodynamics, and for what follows, the reader is referred to any standard text on engineering thermodynamics, e.g. Wallace and Linning† (1968).

† Wallace, F. J. *and* Linning, W. A. (1968). *Basic engineering thermodynamics.* Sir Isaac Pitman and Son Ltd., London.

Thermal efficiency

The ratio of the work produced W to the energy supplied Q is called the thermal efficiency η, so that $\eta = W/Q$. In many applications, it is important to maximize the thermal efficiency, since this represents the fraction of 'useful' energy obtained from that energy which is purchased in the form of gallons of petrol or oil. It is of interest, therefore, to establish the maximum possible value of thermal efficiency, bearing in mind the limitation of the second Law of Thermodynamics that it must always be less than unity.

Carnot efficiency

For any given situation, the theoretical maximum thermal efficiency depends only on the maximum and minimum temperature of the cycle, and is given by

$$\eta_{max} = (T_{max} - T_{min})/T_{max}.$$

This relationship is so important that it is given the special name 'Carnot efficiency'. It is the highest possible value, and is attained when all heat transfers to, or from, the system occur at the constant temperatures of T_{max} or T_{min} respectively.

P–V and T–S diagrams

The processes which occur in the simplest thermal machine are still, however, so complicated that it is not possible to precisely calculate what is happening. Instead, a theoretical model is assumed, in which the various events are idealized to the extent necessary to make analysis of their operation possible. In this way, the operation of most types of machines may be simulated by the assumption of a repeated sequence of thermodynamic processes, called a *cycle*. Usually, each process is assumed to be one in which changes in the thermodynamic functions are occurring as the fluid moves from one state to another, but one of the functions is maintained constant. The important thermodynamic functions here are pressure (P), volume (V), temperature (T), internal energy (U), enthalpy (W) and entropy (S).

A cycle, consisting of a sequence of processes in which one of the thermodynamic functions is maintained constant while the others change, can be graphically represented in a variety of ways. Two of these are of importance in aiding the analysis of the operation of thermal machines. These are the pressure–volume (P–V) and the temperature–entropy (T–S) diagrams.

These two diagrams are important because *areas on the P–V diagram represent work done* and *areas on the T–S diagram represent heat transferred*. As an example, consider Fig. 2.1, which shows a piston in a closed-ended cylinder. Some gas is trapped in the volume contained between the end of the cylinder and the piston, and can be said to be at a state represented by the point A, shown on the pressure–volume and temperature–enthalpy planes. If this gas was now heated through the cylinder walls, from some external source, a number of different things might happen. If the piston was fixed, the volume would remain constant, and heating the gas would result in increases in the pressure and temperature, as shown in Fig. 2.1(a). The supplied heat would be the shaded

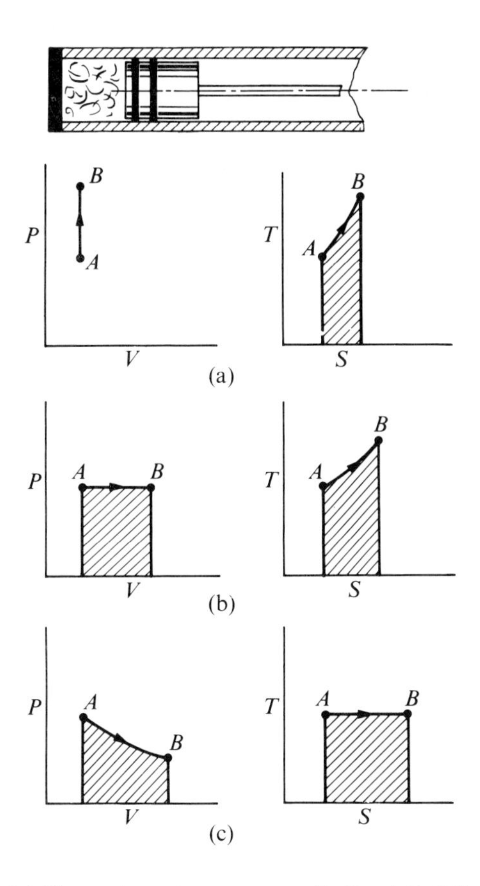

Fig. 2.1 Thermodynamic processes on $P-V$ and $T-S$ diagrams.
(a) Constant-volume heating.
(b) Constant-pressure heating.
(c) Constant-temperature heating.

area shown on the $T-S$ diagram. Any work done would be the area of the $P-V$ diagram: in this case there is no change in volume so no work is done. If, instead, the piston was free to move, and the process of heat addition was regulated, so as to maintain the pressure or temperature constant, the $P-V$ and $T-S$ diagrams, shown in Fig. 2.1(b) and 2.1(c), would result. In both these cases, work is *done* by the gas in expanding to a larger volume as the heat is added to the system.

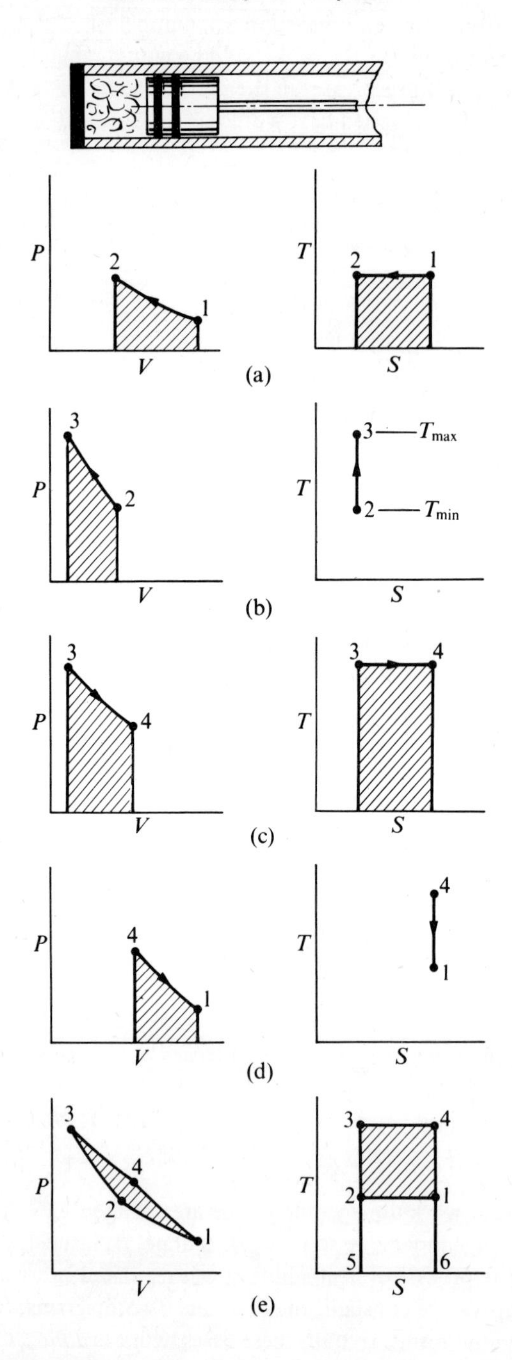

Fig. 2.2. The Carnot cycle.
(a) Isothermal compression.
(b) Isentropic compression.
(c) Isothermal expansion.
(d) Isentropic expansion.
(e) Combined diagrams for complete Carnot cycle.

The Carnot cycle

The Carnot cycle is a thermodynamic cycle comprised of four processes, occurring sequentially, as shown in Fig. 2.2.

To consider the operation of an ideal Carnot-cycle engine, let us assume that we have a cylinder and piston, as shown in Fig. 2.2. We assume that the cylinder is perfectly insulated, and that the piston can move, with no friction and no leakage of the working fluid from the cylinder. The cylinder head is a component that can be perfectly conducting or perfectly insulating, as we choose.

For the start of the cycle we will assume the piston to be at the *outer dead point* (O.D.P.), so that the volume contained within the piston and cylinder combination is a maximum. The pressure and temperature (T_{min}) of the working fluid are at their minimum values, and are represented on Fig. 2.2(a) by the point 1. We let the piston move towards the cylinder head, so that compression occurs, shown by the process 1–2 on Fig. 2.2(a). For this process, we assume that the cylinder head is perfectly conducting, and that the heat-transfer rate is infinite, so that the process occurs isothermally (constant temperature). Work is done *on* the gas, represented by the shaded area on the P–V diagram; heat is *abstracted* from the working fluid, represented by the shaded area on the T–S diagram. In this case, since the process is isothermal, the amount of the heat transferred is exactly equal to the work done (in comparable units).

For the second process, isentropic compression, shown in Fig. 2.2(b), the cylinder head is made perfectly insulating. As the pistons continue to move towards the cylinder head, heat can no longer be abstracted from the working fluid, and so, ideally, the entropy remains constant. This process results in a decrease in the volume, and in increases in both the pressure and temperature. The work done on the gas is the shaded area on the P–V diagram, but there is no heat transferred. The remaining two processes, isothermal expansion from 3 to 4 and isentropic expansion from 4 to 1 then follow, and are shown on Fig. 2.2(c) and Fig. 2.2(d), respectively.

If these four diagrams are combined, the resultant P–V and T–S diagrams are as shown in Fig. 2.2(f). The shaded area, enclosed by the envelope 1–2–3–4 on the P–V diagram, is the useful work produced by the cycle. Similarly, on the T–S diagram, the area 1–2–5–6 is the heat supplied to the cycle. The area 1–2–3–4 is the amount converted to work, and the area 1–2–5–6 is the 'waste heat' of the cycle. It is clear, from this diagram, why the Carnot cycle has the highest possible thermal efficiency. Given temperature limits, T_{max} and T_{min}, no possible sequence of thermodynamic processes could result in a larger ratio of the areas 1–2–3–4 and 1–2–5–6, so that the efficiency, $\eta = W/Q$ = area 1–2–3–4/ area 1–2–5–6 must be a maximum.

Absolute temperatures must be used in thermodynamic analysis. The zero temperature on the T–S diagram is −273 °C (= 0° K) or −460°F (= 0°R), so that the 'waste-heat' area 1–2–5–6 may be very appreciable.

It is clear that the efficiency of the Carnot cycle (and this generally applies to all engines) can be improved by (a) increasing T_{max} and (b) decreasing T_{min}. The ultimate maximum value of T_{max} is governed by the materials used to construct the engine, this is called the 'metallurgical limit'. The lowest possible value of T_{min} is that temperature at which cooling water or air is available, generally, the ambient atmospheric temperature.

In practice, it is not possible to construct Carnot-cycle engines. There are no materials which are perfectly insulating or conducting, and all pistons sliding in cylinders do have friction and leakage losses. However, the most serious difficulty arises because isothermal and isentropic processes for a gas (say, air), have slopes that are so little different, when compared on a $P-V$ diagram, that the area of the $P-V$ diagram shown in Fig. 2.2(e) becomes negligibly small, unless pressures of millions of pounds per square inch, and piston strokes of several feet, are used. This would result in a tremendously heavy engine, which would be quite unable to produce sufficient work to overcome its own friction losses. Despite this lack of practicality, the Carnot cycle is useful in a preliminary study of the operation of an engine. Furthermore, with some modifications (which change it to the Rankine cycle), the Carnot cycle is representative of the mode of operation of liquid-vapour machines, such as reciprocating steam-engines, steam turbines, or 'Freon' refrigerating plants.

The Stirling cycle

The Stirling cycle is similar, in some respects, to the Carnot cycle. It is illustrated in Fig. 2.3.

Consider a cylinder containing two opposed pistons, with a regenerator between the pistons. The regenerator may be thought of as a thermodynamic sponge, alternately releasing and absorbing heat. It is a matrix of finely-divided metal in the form of wires or strips. One of the two volumes between the regenerator and the pistons is called the *expansion space*, and is maintained at a high temperature T_{max}. The other volume is called the *compression space*, and is maintained at a low temperature T_{min}. There is, therefore, a temperature gradient $(T_{max} - T_{min})$ across the transverse faces of the regenerator, and it is assumed that there is no thermal conduction in the longitudinal direction. As in the Carnot cycle, it is assumed that the pistons move without friction or leakage loss of the working fluid enclosed between them.

To start the cycle, we assume that the compression-space piston is at the outer dead point, and the expansion-space piston is at the inner dead point, close to the face of the regenerator. All the working fluid is then in the cold compression space. The volume is a maximum, so that the pressure and temperature are at their minimum values, represented by 1 on the $P-V$ and $T-S$ diagrams, shown in Fig. 2.3. During compression (process 1–2), the compression piston moves towards the inner dead point, and the expansion-space piston remains stationary. The working fluid is compressed in the compression space, and the pressure

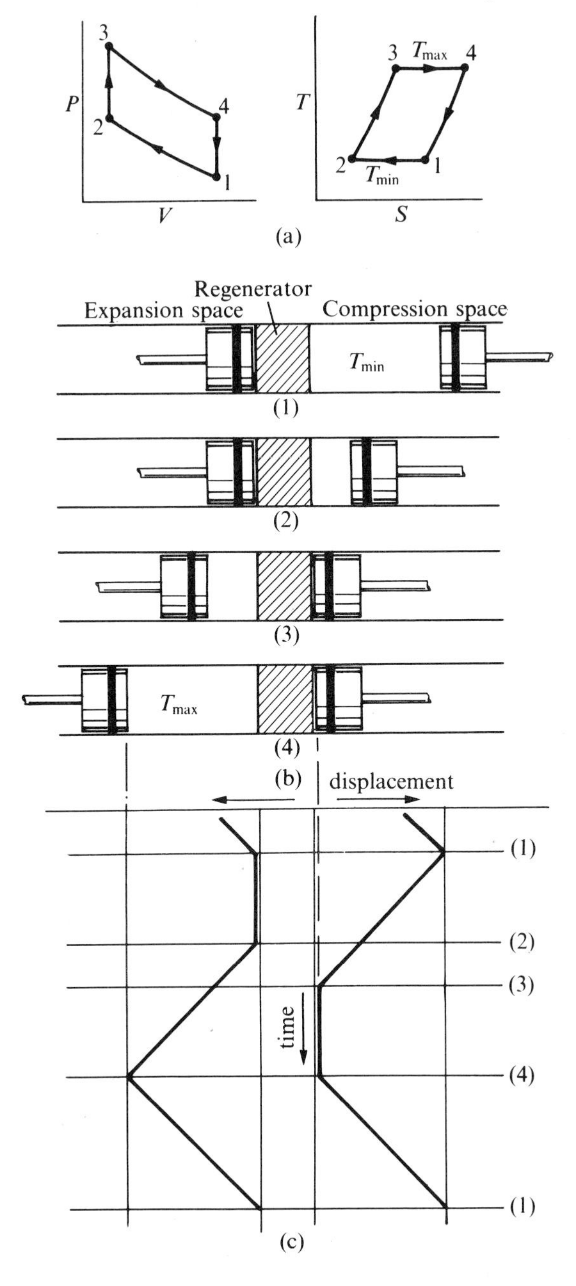

Fig. 2.3. The Stirling cycle.
(a) $P-V$ and $T-S$ diagrams.
(b) Piston arrangement at the terminal points of the cycle.
(c) Time–displacement diagram.

increases. The temperature is maintained constant because heat Q_c is abstracted from the compression-sapce cylinder to the surrounds.

In the transfer process 2–3, both pistons move simultaneously, the compression piston towards (and the expansion piston away from) the regenerator, so that the volume between them remains constant. Therefore, the working fluid is transferred, through the porous metallic matrix of the regenerator, from the compression space to the expansion space. In passage through the regenerator, the working fluid is heated from T_{min} to T_{max}, by heat transfer from the matrix, and emerges from the regenerator into the expansion space at temperature T_{max}. The gradual increase in temperature in passage through the matrix, at constant volume, causes an increase in pressure.

In the expansion process 3–4, the expansion piston continues to move away from the regenerator towards the outer dead point; the compression piston remains stationary at the inner dead point, adjacent to the regenerator. As the expansion proceeds, the pressure decreases as the volume increases. The temperature remains constant because heat Q_E is added to the system from an external source.

The final process in the cycle is the transfer process 4–1, during which both pistons move simultaneously to transfer the working fluid (at constant volume) back, through the regenerative matrix form and the expansion space, to the compression space. In passage through the matrix, heat is transferred from the working fluid to the matrix, so that the working fluid decreases in temperature, and emerges at T_{min} into the compression space. Heat transferred in the process is contained in the matrix, for transfer to the gas in process 2–3 of the subsequent cycle.

The cycle is composed, therefore, of four heat-transfer processes.

Process 1–2– isothermal compression; heat transfer *from* the working fluid at T_{min} to the external dump.

Process 2–3– constant volume; heat transfer *to* the working fluid from the regenerative matrix.

Process 3–4– isothermal expansion; heat transfer *to* the working fluid at T_{max} from an external source.

Process 4–1– constant volume; heat transfer *from* the working fluid to the regenerative matrix.

If the heat transferred in process 2–3 has the same magnitude as in process 4–1, then the only heat transfers between the engine and its surroundings are (a) heat supply at T_{max} and (b) heat rejection at T_{min}. This heat supply and heat rejection at constant temperature satisfies the requirement of the second Law of Thermodynamics for maximum thermal efficiency, so that the efficiency of the Stirling cycle is the same as the Carnot cycle, i.e. $\eta = (T_{max} - T_{min})/T_{max}$. The principal advantage of the Stirling cycle over the Carnot cycle lies in the replacement of two isentropic processes by two constant volume processes, which greatly increases the area of the P–V diagram. Therefore, to obtain a reasonable amount of work from the Stirling cycle, it is not necessary to resort to very high pressures and swept-volumes, as in the Carnot cycle.

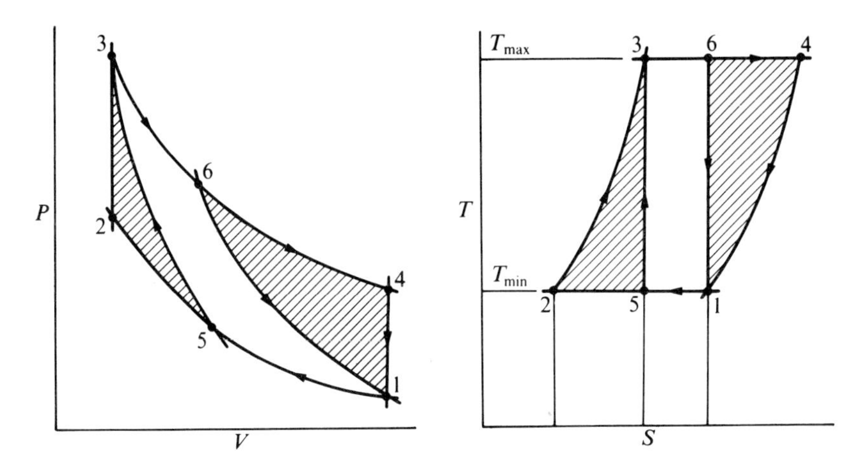

Fig. 2.4. Stirling and Carnot cycles. The Stirling and Carnot cycles are superimposed, with common values for the maximum and minimum temperatures, pressures, and volumes. Hatched areas on the P–V plane represent the increased work output of the Stirling cycle. Hatched areas on the T–S plane represent increased heat-transfer of the Stirling cycle.

A comparison of the P–V diagrams of a Carnot and Stirling cycle, between given limits of pressure, volume, and temperature, is shown on Fig. 2.4. The shaded areas 5–2–3 and 1–6–4 represent the additional work made available by substituting constant-volume processes for isentropic processes. The isothermal processes (1–5 and 3–6) of the Carnot cycle are extended to process 1–2 and 3–4, respectively, so that the quantities of heat supplied to – and rejected from – the Stirling cycle are increased in the same proportion as the available work. The *fraction* of supplied heat which is converted to work (the efficiency), is the same in both cycles.

The Ericsson cycle

In the Ericsson cycle, the processes of constant-volume regenerative heat transfer, described above, are replaced by constant-pressure regenerative processes. This leads to the P–V and T–S diagrams shown in Fig. 2.5. The efficiency of the cycle is the same as that of the Carnot cycle but, as in the Stirling cycle, the net available work and the quantities of heat transferred are much greater, for given limits of pressure, volume, and temperature.

The Stirling cycle as a prime mover

In the previous discussion, heat, at some high temperature T_{max}, was supplied to the cycle. Part of the heat was converted to work, and part was rejected, as heat, at a low temperature T_{min}. This describes a cycle operating as a prime mover, a machine producing work from a high-temperature energy source, and rejecting heat at a low temperature.

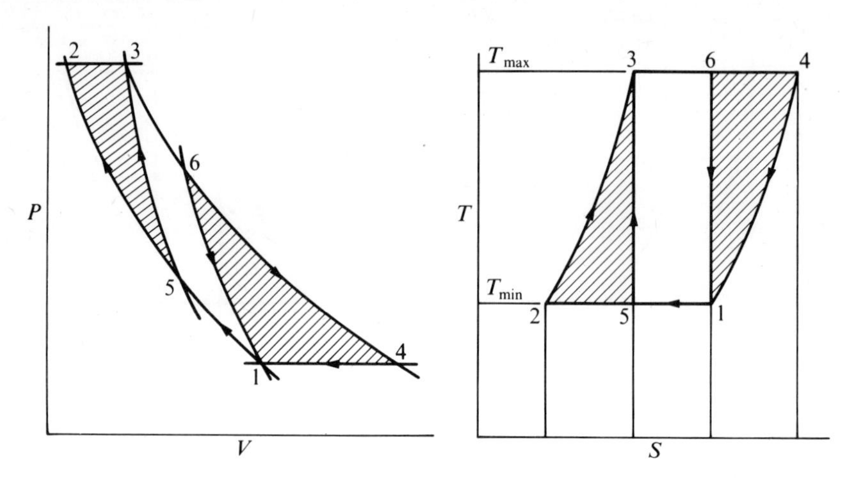

Fig. 2.5. Ericsson and Carnot cycle. The Ericsson and Carnot cycles are superimposed, with common values for the maximum and minimum temperatures, pressures, and volumes. As before, the hatched areas represent the increased work-output and heat-transfer of the Ericsson cycle.

The Stirling cycle as a refrigerating machine

The same ideal machine which was used to describe the operation of the Stirling cycle as a prime mover can be used to describe the operation of the cycle as a refrigerating machine. The only difference is that the temperature of the heat supplied from the external source during expansion is *lower* than the temperature at which heat is rejected from the working fluid during compression. This is illustrated in Fig. 2.6, where P–V and T–S diagrams for the prime mover and refrigerating machine are superimposed.

When the Stirling-cycle machine is operating as a refrigerator, heat is lifted from the cold zone during the expansion process 3′–4′. The work of compression (area 1–2–5–6) is the same for both the prime mover and refrigerator. The work of expansion (area 4′–3′–5–6), in the case of a refrigerator is less than the compression work and work equivalent to area 1–2–3′–4′, from an external source, is necessary to drive the cycle. During transfer from the compression space to the expansion space, in process 2–3′, the working fluid experiences a *decrease* in temperature, and a corresponding increase in temperature during the alternate transfer process 4′–1.

The performance of a refrigerator is assessed in terms of its coefficient of performance (COP), where

$$\text{COP} = \text{heat lifted/work done} = T_{ref}/(T_{min} - T_{ref}).$$

The COP of the Stirling, Ericsson, and Carnot cycles are the same, for given temperature limits, but the *refrigerating* capacity of the Stirling and Ericsson cycles are much greater than the Carnot cycle, for given pressure and volume limits.

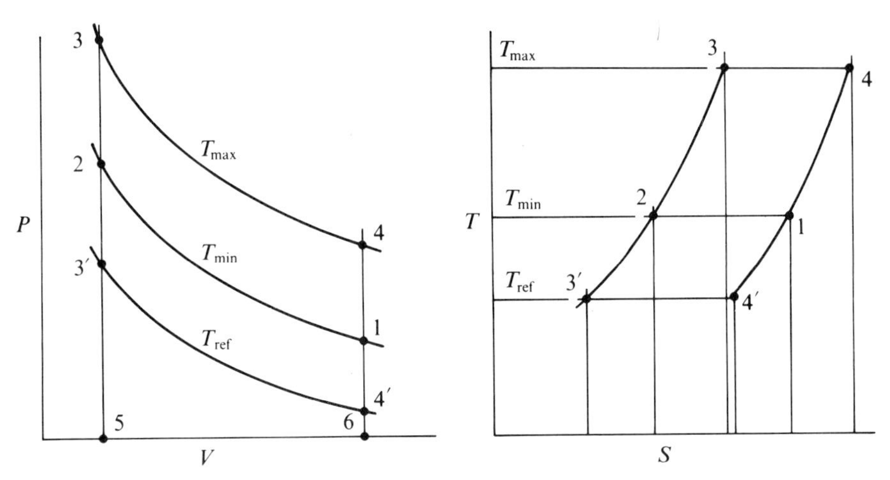

Fig. 2.6. Stirling cycle as prime mover and as cooling engine. In both prime movers and cooling engines the *compression* process occurs at temperature T_{min}. The *expansion* process occurs at a temperature of (a) T_{max} in the prime mover, and (b) T_{ref} in the cooling engine. Heat is supplied at a high temperature to produce useful work in the prime mover. Heat is abstracted to refrigerate, in the cooling engine, and a net input of work is required.

The Stirling cycle as a heat pump

As a heat pump, the Stirling cycle operates exactly as it did in the refrigerating machine described above, with the temperature of the expansion space T_{ref} less than the temperature of the compression space T_{min}. The difference between operation as a heat pump and refrigerating machine is that both T_{ref} and T_{min} are increased. In both the prime mover and refrigerator application, T_{min} is the ambient atmospheric temperature at which cooling water is available, whereas, in the case of the heat pump, T_{min} is the temperature at which heat is rejected from the system, and is the useful product, for heating a concert hall or office building. Therefore, for a heat pump, T_{min} is *above* the ambient atmospheric temperature, and heat is supplied to the cycle (at T_{ref}), from atmospheric air or river water, at approximately the ambient atmospheric temperature.

A comparison of the Stirling cycle's performance as a heat pump and a refrigerator is drawn in Fig. 2.7. In both cases, work from an external source is required to drive the cycle, and is equivalent to area 1–2–3–4. In the case of the heat pump, the useful product is the heat rejected at temperature T_{min}, and the performance of a heat pump is therefore assessed as

$$COP_{HP} = (\text{heat rejected})/(\text{work done}) = (T_{min})/(T_{min} - T_{ref}).$$

This is the inverse of the thermal efficiency, whereas the coefficient of performance of a refrigerator, namely,

$$\mathrm{COP}_{\mathbf{ref}} = (T_{\mathbf{ref}})/(T_{\mathbf{min}} - T_{\mathbf{ref}}),$$

is *not* the inverse of thermal efficiency.

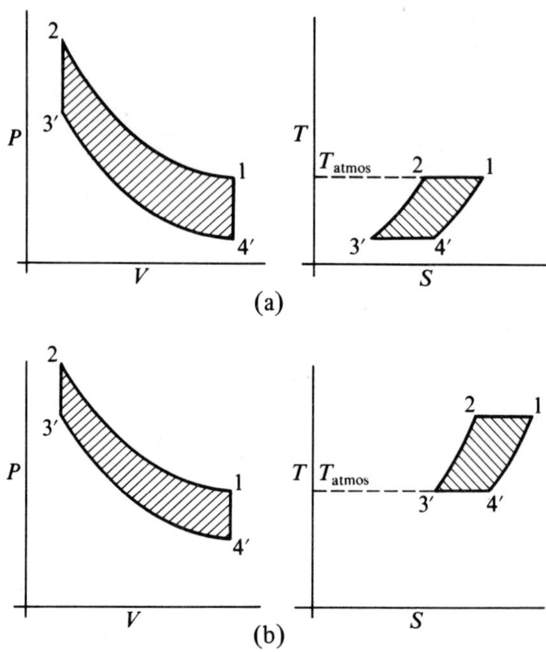

Fig. 2.7. Stirling cycle as a heat pump and a refrigerating machine. (a) Heat pump. (b) Refrigerating machine. When operating as a heat pump or refrigerating machine, the expansion process occurs at a temperature less than that for the compression process, and a net input of work is required. In the heat pump, expansion is at atmospheric temperature, and the heat rejected during compression, at a high temperature, is the useful output of the cycle. In the refrigerating machine, the heat supplied, during expansion at a low temperature, is the useful refrigerating capacity of the cycle.

The Stirling cycle as a pressure generator

Systems closely related to the Stirling cycle have been proposed, and are under investigation, where the objective is to pump a fluid, and to increase the pressure. When the fluid to be pumped is a liquid (or gas), separated by a diaphragm (or piston) from the working fluid in the Stirling cycle device, the system can be classified as a Stirling cycle which is working as a prime mover. In other instances, where the working fluid is itself the fluid to be compressed and pumped, there are, invariably, valves or other flow-controlling devices; these systems cannot be classified as Stirling-cycle machines, using the limited definition proposed in the introduction. Nevertheless, they are discussed in the literature as Stirling-cycle machines. In most cases, fluid is added to the cycle when the pressure is low, and withdrawn at a higher pressure. Heat is supplied at a high temperature, and rejected at a low temperature. The work during expansion is greater than the work during compression, by an amount equivalent to the 'pump' work of the compressed fluid.

Conclusion

Most introductions to applied thermodynamics emphasize the singularity of the Carnot cycle as an idealized system with no practical application, but with the highest possible thermodynamic efficiency. It is not generally understood that there are an infinite number of thermodynamic cycles that have the same maximum thermal efficiency. All these cycles have some form of regenerative transfer processes and isothermal heat addition and rejection. The one exception is the Carnot cycle, which has *isentropic* processes instead of *regenerative* processes. Two other special cases are the Stirling cycle and the Ericsson cycle, where the regenerative processes are at constant volume and constant pressure, respectively. An infinite number of other cycles may be postulated in which the regenerative processes are neither at constant volume, at constant pressure, nor isentropic, but none of these have particular names.

The possibility of constructing a machine to operate on the Stirling cycle is no less remote than the possibility of constructing a Carnot engine. Machines that are now called Stirling engines do not, in fact, operate on the Stirling cycle. The differences between the ideal cycle and the operation of practical engines are considered in the following chapter.

3 Practical regenerative engine cycle

Ideal cycle

The Stirling cycle is a highly-idealized thermodynamic cycle, comprised of four thermodynamic processes, including two isothermal and two constant-volume processes.

In previous discussion of the cycle, it was assumed that all processes were thermodynamically reversible, and that the processes of compression and expansion were isothermal, thereby implying infinite rates of heat transfer between the cylinder walls and working fluid. It was further assumed that all the working fluid was in the compression or expansion space during the processes of expansion and compression, so that the effects of any voids in the regenerative matrix, clearance space, or pockets in the cylinder were neglected. The two pistons were caused to move in some discontinuous fashion to achieve the prescribed working-fluid distribution, and all aerodynamic- and mechanical-friction effects were neglected. Regeneration was assumed to be perfect, which implied an infinite rate of heat transfer between the working fluid and regenerative matrix, and an infinite heat capacity of the regenerative matrix.

Practical cycle

In any practical engine, all these factors and others combine to reduce the thermal efficiency to well below the Carnot value of the ideal cycle. The actual thermal efficiency may be quoted as a fraction of the theoretical Carnot efficiency; this ratio is called the relative efficiency,

$$\eta_{rel} = \text{actual thermal efficiency/Carnot thermal efficiency.}$$

A value in excess of 0·4 for the relative efficiency is evidence of a well-designed machine.

To illustrate the discussion of the ideal cycle, a mechanical arrangement was assumed of two opposed pistons, with an interposed regenerator. The two-piston machine is one of several different mechanical arrangements which are to be considered in detail later. One practical version of a two-piston machine is shown in Fig. 3.1. It consists of a **V** engine, with both pistons coupled to a common crankshaft. The spaces above the pistons constitute the compression and expansion volumes; they are coupled by a duct, containing the regenerator and additional heat-exchangers.

In the operation of this engine, a significant departure from ideality arises as a

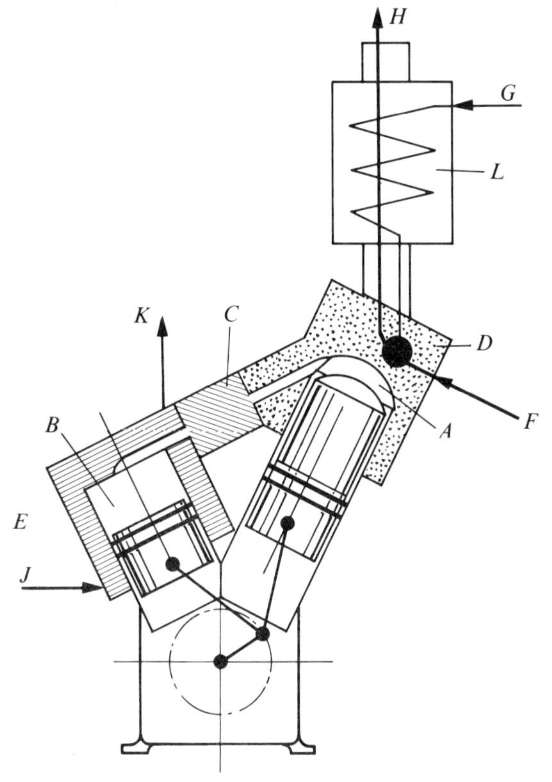

Fig. 3.1. Diagram of practical opposed-piston Stirling engine.
A – expansion space, B – compression space, C – regenerator, D – heater, E – cooler, F – fuel inlet, G – air inlet, H– exhaust products of combustion, J – water inlet, K water outlet, L – exhaust gas inlet–air preheater.

consequence of the continuous, rather than discontinuous, motion of the pistons. This results (as shown in Fig. 3.2) in a P–V diagram which is a smooth continuous envelope. The four processes of the ideal cycle are not sharply defined.

The processes of compression and expansion do not take place wholly in one or other of the two spaces, so that three P–V diagrams may be drawn, one for the compression space, one for the expansion space, and one for the total enclosed volume, which includes the 'dead' space. The 'dead' space is defined as that part of the working space not swept by one of the pistons, and includes cylinder clearance spaces, void volumes of the regenerator and other heat-exchangers, and the internal volume of associated ducts and ports. The P–V diagram for the expansion space represents the total positive work of the cycle, whereas the diagram for the compression space represents the compression (or negative) work of the cycle. The difference in the areas of these diagrams is the net cycle-output, the 'indicated' work available for overcoming mechanical-friction losses and for providing useful power to the engine crankshaft.

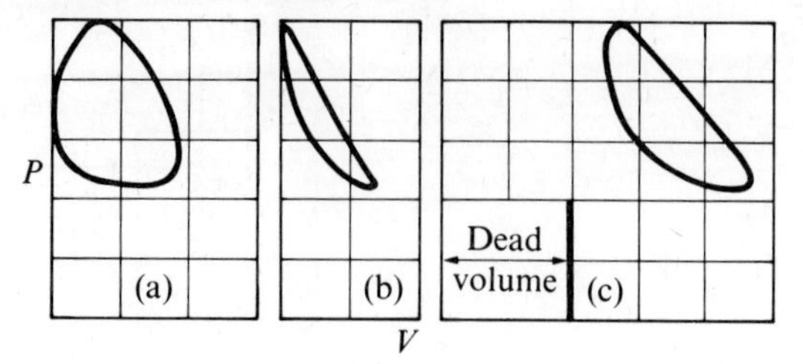

Fig. 3.2. Pressure–volume diagrams for practical engine.
(a) Expansion-space diagram.
(b) Compression-space diagram.
(c) Total working-space.

In a cycle where the processes of compression and expansion are isothermal and there are no friction losses, the difference in the area of the expansion- and compression-space diagrams will be found to be exactly equal to the area of the P–V diagram for the total working space. In a practical engine, of course, this equality does not obtain, because aerodynamic-flow losses in the regenerator and other heat-exchangers cause differences in the pressure of the working fluid in the compression and expansion spaces. Flow losses are important, because (as shown in Fig. 3.3), they cause a decrease in the area of the expansion-space P–V diagram, resulting in (a) a decrease in the net cycle-output (and, hence, in efficiency) of a prime mover and (b) a decrease in the cooling capacity and the COP of a refrigerating machine.

The sinusoidal piston motion results in the working fluid being distributed in a cyclically time-variant manner throughout various temperature ranges, and it is not possible to draw a meaningful T–S diagram for the total mass of the working fluid. It is possible to draw T–S diagrams for particular particles of the working fluid, as they move from one temperature range to another, but no convenient way has been found to combine these multiple diagrams.

The processes of compression and expansion are not isothermal, another major departure from ideality. In an engine, running at a reasonable speed (say, 1000 rev/min), it is likely that the processes are nearer adiabatic (no heat-transfer) than isothermal (infinite heat-transfer). In order to improve the situation special heat-exchangers are often incorporated (as shown in Fig. 3.1), including (a) a heater, adjacent to the expansion space, imparting heat *to* the working fluid, and (b) a cooler, adjacent to the compression space, abstracting heat *from* the working fluid. Despite the advantages of improved heat-transfer, the provision of such heat-exchangers imposes some penalties. Additional aerodynamic-flow losses are likely, with the consequent deleterious effect on performance, as discussed above. The dead volume will be increased by the void volume of the heater and cooler, and this has a critical effect on the performance of regenerative engines. Furthermore, the working fluid is heated, not only when flowing from the regenerator *to* the expansion space, but also when flowing *from* the expansion space to the regenerator. Similarly, the working fluid is cooled when flowing to, as well as from, the compression space. The provision of one-way flow systems is possible, but adds much complication to the machine.

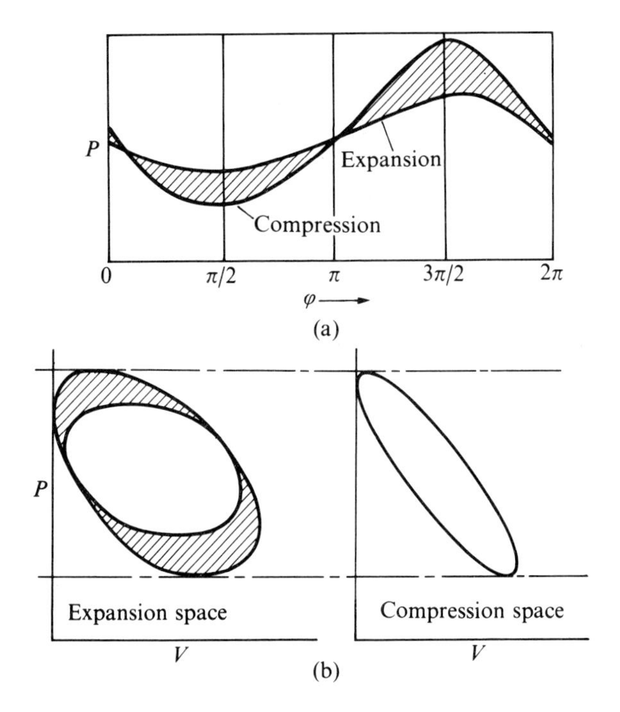

Fig. 3.3. Effect of aerodynamic-flow loss on engine work.
(a) Pressure–time diagram for pressure variation in the expansion and compression spaces. The difference in pressure is the flow-loss in the regenerator and other heat-exchangers.
(b) Pressure–volume diagrams for the compression and expansion spaces. The hatched area on the diagram for the expansion space represents the work effectively lost by flow-losses in the regenerator and other heat-exchangers.

Considerations of increased flow-loss and void-volume (along with considerations of cost, size, and weight) combine to produce a compromised heat-exchanger design. Consequently, substantial differences may exist between the temperatures at which heating (combustion products) and cooling (water or air) is available and the temperatures experienced by the working fluid. This is illustrated diagrammatically in Fig. 3.4, and might be considered representative of the temperature range established in a fossil-fuelled water-cooled regenerative engine. The temperatures of the combustion products and cooling water are 2800K and 280K, respectively. The metallurgical limit of the materials used for the expansion cylinder and heater is (say), 1000K. This provides for a steep temperature gradient of 2800 to 1000K between the combustion products and cylinder wall, with the potential for high rates of heat transfer. Further temperature gradients of (say), 100K, between the working fluid and the expansion space, and 50K, between the working fluid and the compression space, might exist, so that the cyclic temperature excursion of the working fluid varies from (280 + 50) = 330K to (1000 − 100) = 900K. Whereas the Carnot- (or Stirling-) cycle efficiency for the system *might* be calculated as

$$\eta_c = (2800 - 280)/2800 = 2520/2800 = 90 \text{ per cent},$$

to give a more realistic picture it *should* be calculated as

$$\eta_c = (900 - 330)/900 = 570/900 = 63 \text{ per cent}.$$

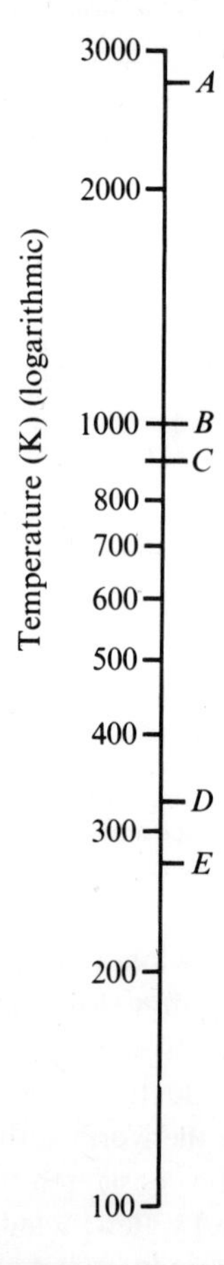

Fig. 3.4. Characteristic temperature regime in a fossil-fuelled, water-cooled Stirling engine.
A – Temperature of combustion products,
B – Temperature of heater walls,
C – Mean temperature in expansion space,
D – Mean temperature in compression space,
E – Temperature of cooling water and cooler walls.

This example demonstrates one of the major difficulties in the commercial application of Stirling engines – one shared by the gas turbine – the question of materials. Some parts of the machine (the heater and expansion space), are exposed, continuously, to a high temperature, and are subject, therefore, to the metallurgical limit of the heater and expansion cyclinder materials.

The allowable temperature-excursion of the working fluid in a Stirling engine is limited to a fraction of that permissible in an internal-combustion, Otto- or Diesel- cycle engine, where the maximum cycle temperatures are attained only momentarily. Thus, although regenerative cycles between given temperature limits are thermodynamically more efficient than Otto- or Diesel-cycles, in practice regenerative engines are compared with gas (or oil) engines operating with radically different temperature limits.

Not all the heat available from combustion of the fuel and air can be transferred to the working fluid, since this would require a very large heater. The heat passing to exhaust in the combustion products of a Stirling engine represents a direct loss, because it must be paid for in terms of gallons of oil (or cubic feet of gas burned), but has served no useful purpose in the engine. An important engine accessory, therefore, is another heat-exchanger (the exhaust–air preheater), used to warm the incoming air by heat transferred from the exhuast gas. This heat-exchanger can be of the recuperative type or the regenerative type. In the recuperative type, the two fluids, exhaust gas, and incoming air are separated by walls into separate ducts. In the regenerative type, the fluids flow alternately, and usually in contraflow, through the same porous matrix. It is important to distinguish carefully between the regenerative heat-exchanger, incorporated as an integral part of the engine, and the recuperative (or regenerative) heat-exchanger, used as an accessory of the engine for exhaust–air preheating.

The continuous motion of the reciprocating elements, the non-isothermal compression and expansion processes, the limited heat-transfer in cooling and heating devices, the exhaust-stack loss, the increased dead space, and aerodynamic-flow loss together constitute the principal reasons for the failure of most practical Stirling engines to fulfil their designers hopes and ambitions. Other causes of disappointment include deficiencies in regenerator operation, high mechanical-friction losses, temperature equalization as a result of relatively massive conduction paths, and fluid leakage owing to imperfectly designed (or imperfectly operating) seals.

4 Theoretical analysis of Stirling cycle systems

Ideal Stirling cycle

Equations for analysis of the ideal Stirling cycle are summarized below. The gross idealization of the Stirling cycle precludes the use of these equations for anything other than elementary preliminary calculations.

With reference to Fig. 2.3 and the discussion of the ideal cycle given in Chapter 2, the equations may be summarized as follows.

Required data

(i) Some reference temperature and pressure, or volume, say, conditions, at state 1.

(ii) Temperature ratio $\tau = T_{min}/T_{max}$.

(iii) Volume ratio $r = V_{max}/V_{min}$.

For unit mass of working fluid, assumed to be a perfect gas, then $V_1 = RT_1/p_1$, from the characteristic gas equation.

The following data is required for each of the four cycle processes.

(a) *Isothermal compression process* (1–2). In this process, heat is abstracted from the working fluid and rejected from the cycle at the maximum cycle temperature. Work is done on the working fluid equal in magnitude to the heat rejected from the cycle. There is no change in internal energy, and there is a decrease in entropy.

$$p_2 = p_1 V_1/V_2 = p_1 r;\ T_2 = T_1 = T_{min}.$$

Heat transfer (Q) = work done $(W) = P_1 V_1 \ln(1/r) = RT_1 \ln(1/r)$.
Change in entropy $(S_2 - S_1) = R\ln(1/r)$.

(b) *Constant-volume regenerative transfer process* (2–3). In this process heat is transferred from the regenerative matrix to the working fluid, increasing the temperature from T_{min} to T_{max}. No work is done, and there is an increase in the entropy, and internal energy, of the working fluid.

$$p_3 = p_2 T_3/T_2 = p_2/\tau,\ V_3 = V_2.$$

Heat transfer $(Q) = C_v(T_3 - T_2)$.
Work done $(W) = 0$.
Change in entropy $(S_3 - S_2) = C_v \ln(1/\tau)$.

(c) *Isothermal expansion process* (3–4). In this process, heat is supplied to the cycle at a high temperature T_{max}, during expansion of the working fluid. Work is done, by the working fluid, equal in magnitude to the heat supplied. There is no change in the internal energy, but an increase in the entropy of the working fluid.

$$p_4 = p_3 V_3/V_4 = p_3(1/r);\ T_4 = T_3 = T_{max}.$$

Heat transfer (Q) = work done (W) = $p_3 V_3 \ln r = RT_3 \ln r$.
Change in entropy $(S_4 - S_3) = R \ln r$.

(d) *Constant-volume regenerative transfer process* (4–1). In this process, heat is transferred from the working fluid to the regenerative matrix, decreasing the temperature of the working fluid from T_{max} to T_{min}. No work is done, and there is a decrease in the internal energy and entropy of the working fluid.

$$p_1 = p_4 T_4/T_1 = p_1 \tau,\ V_1 = V_4$$

Heat transfer (Q) = $C_v(T_1 - T_4)$.
Change in entropy $(S_1 - S_4) = C_v \ln \tau$.

In the regenerative processes, the heat transferred from the matrix to the working fluid in process 2–3 is restored from the working fluid to the matrix in process 4–1. There is no net gain or loss of heat by the working fluid or the matrix. Therefore

the total heat supplied (at T_{max}) = $RT_3 \ln r$,
the total heat rejected (at T_{min}) = $RT_1 \ln(1/r)$,

$$\text{and thermal efficiency} = \frac{\text{heat supplied} - \text{heat rejected}}{\text{heat supplied}}$$

$$= \frac{\text{work done}}{\text{heat supplied}}$$

$$= (RT_3 \ln r - RT_1 \ln r)/RT_3 \ln r$$

$$= 1 - \tau.$$

This value corresponds to the Carnot efficiency between the same temperature limits.

The Schmidt cycle

The classical analysis of the operation of Stirling engines is due to Schmidt (1861). The theory provides for harmonic motion of the reciprocating elements, but retains the major assumptions of isothermal compression and expansion and of perfect regeneration. It, thus, remains highly idealized, but is certainly more realistic than the ideal Stirling cycle. Provided a reasonable level of caution is exercised in interpretation, the predictions of the Schmidt theory can be a useful tool for engine design.

Attempts to improve realism, by modifying the assumptions of isothermal processes and perfect regeneration, have introduced considerable complexity and resulted in non-closed-form solutions requiring the application of digital, or analog, computer simulations. Detailed discussion of these are beyond the scope of this book. Experience suggests that a limited development effort is best invested in the use of a reliable, relatively simple, idealized theory, with subsequent engineering development of the hardware. Computer simulation of engines and design-optimization thereof appears justified only in advanced research and development programmes, or as a component of academic thesis projects.

Principal assumptions of the Schmidt cycle

1. The regenerative process is perfect.
2. The instantaneous pressure is the same throughout the system.
3. The working fluid obeys the characteristic gas equation, $PV = RT$.
4. There is no leakage, and the mass of working fluid remains constant.
5. The volume variations in the working space occur sinusoidally.
6. There are no temperature gradients in the heat-exchangers.
7. The cylinder wall, and piston, temperatures are constant.
8. There is perfect mixing of the cylinder contents.
9. The temperature of the working fluid in the ancillary spaces is constant.
10. The speed of the machine is constant.
11. Steady state conditions are established.

Nomenclature used in the following analysis†

A	= a factor $(\tau^2 + 2\tau\kappa \cos\alpha + \kappa^2)^{\frac{1}{2}}$.
B	= a factor $(\tau + \kappa + 2S)$.
K	= constant.
M	= total mass of working fluid.
N	= machine speed.
p	= instantaneous cycle-pressure.
p_{max}	= maximum cycle-pressure.
p_{mean}	= mean cycle-pressure.
p_{min}	= minimum cycle-pressure.
P	= engine output.

† *Note:* Lower case suffixes indicate instantaneous values of temperature, pressure, volume and mass. Upper case suffices indicate maximum (or constant) values. Thus:
E or e refers to expansion space,
C or c refers to compression space,
D or d refers to dead space.

P_{mass} = P/RT_c dimensionless power parameter based on the mass of working fluid.

P_{max} = $P/(p_{max}V_T)$, dimensionless power parameter, based on the maximum cycle-pressure and combined swept volume.

Q = heat transferred to the working fluid in the expansion space, the heat lifted.

Q_{mass} = Q/RT_c the dimensionless cooling parameter, based on the mass of working fluid.

Q_{max} = $Q/(p_{max}V_T)$, the dimensionless heat lifted, based on the maximum cycle pressure.

R = characteristic gas constant of the working fluid.

S = $(2X\tau)/(\tau + 1)$, reduced dead volume.

T_C = temperature of the working fluid in the compression space, generally assumed to be 300K.

T_D = temperature of the working fluid in the dead space.

T_E = temperature of the working fluid in the expansion space.

V_C = swept volume in the compression space.

V_E = swept volume in the expansion space.

V_D = total internal volume of heat-exchangers, volume of regenerator, and associated ducts and ports.

V_T = $(V_C + V_E) = (1 + \kappa)V_E$, combined swept volume.

V_W = $\frac{1}{2}V_E(1 + \cos\phi) + \frac{1}{2}V_C[1 + \cos(\phi - \alpha)] + V_D$, volume of total working space.

$V_{W\,max}$ = maximum volume of total working space.

X = V_D/V_E, dead-volume ratio.

α = angle by which volume variations in the expansion space lead those in the compression space (in fractions of π radians, or degrees).

δ = $(\tau^2 + \kappa^2 + 2\tau\kappa\cos\alpha)^{\frac{1}{2}}/(\tau + \kappa + 2S)$.

θ = $\tan^{-1}((\kappa\sin\alpha)/(\kappa + \kappa\cos\alpha))$.

κ = V_C/V_E, swept-volume ratio.

τ = T_C/T_E, temperature ratio.

ϕ = crank angle.

Basic equations

Volume of expansion space $\quad V_e = \frac{1}{2}V_E(1 + \cos\phi)$. (4.1)

Volume of compression space $\quad V_c = \frac{1}{2}V_C[1 + \cos(\phi - \alpha)]$ (4.2)

$$= \tfrac{1}{2}\kappa V_E[1 + \cos(\phi - \alpha)]. \quad (4.3)$$

Volume of dead space, being that constant volume of the total working space not included in the volumes of the expansion or compression space,

$$V_D = XV_E. \quad (4.4)$$

Mass of working fluid in expansion space, $M_e = (p_e V_e)/(RT_e)$.
Mass of working fluid in compression space $M_c = (p_c V_c)/(RT_c)$.
Mass of working fluid in dead space $M_d = (p_d V_d)/(RT_d)$.
Since the total mass of the working fluid remains constant,

$$M_T = (p_e V_e)/(RT_e) + (p_c V_c)/(RT_c) + (p_d V_d)/(RT_d) = (KV_E)/(2RT_c). \qquad (4.5)$$

If the instantaneous pressure is the same throughout the system, and equal to p, say, and if T_e and T_c are constant at T_E and T_C then, substituting for the volumes, eliminating R, and re-arranging,

$$K/p = (T_C/T_E)(1 + \cos\phi) + \kappa[1 + \cos(\phi - \alpha)] + (2V_D T_C)/(V_E T_D). \qquad (4.6)$$

If the temperature variation in the dead space is linear in the axial direction, then the mean temperature

$$T_D = T_C + \tfrac{1}{2}(T_E - T_C) = (1 + T_E/T_C)(T_C/2)$$

and, since $T_C/T_E = \tau$ then, from eqn (4.6)

$$K/p = \tau(1 + \cos\phi) + \kappa[1 + \cos(\phi - \alpha)] + 2S \qquad (4.7)$$

where S (the reduced dead volume) $= 2 \times \tau/(\tau + 1)$.
To simplify eqn (4.7) consider first

$$y = x\cos\phi + z\sin\phi, \qquad (4.8)$$

then $y = \sqrt{r^2}\cos(\phi - \beta)$, where $\tan\beta = z/x$, $z = r\sin\beta$, and $x = r\cos\beta$

$$\begin{aligned} \text{since } \sqrt{r^2}\cos(\phi - \beta) &= \sqrt{r^2}(\cos\phi\cos\beta + \sin\phi\sin\beta) \\ &= r\cos\phi\cos\beta + r\sin\phi\sin\beta \\ &= x\cos\phi + z\sin\phi. \end{aligned}$$

Eqn (4.8) is similar in form to eqn (4.7) therefore, by analogy,

$$\begin{aligned} K/p &= [(\tau + \kappa\cos\alpha)^2 + (\kappa\sin\alpha)^2]^{\frac{1}{2}}\cos(\phi - \theta) + \tau + \kappa + 2S \\ &= (\tau^2 + 2\tau\kappa\cos\alpha + \kappa^2)^{\frac{1}{2}}\cos(\phi - \theta) + \tau + \kappa + 2S, \end{aligned} \qquad (4.9)$$

where $\tan\theta = (\kappa\sin\alpha)/(\tau + \kappa\cos\alpha)$

Let $A = (\tau^2 + 2\tau\kappa\cos\alpha + \kappa^2)^{\frac{1}{2}}$, $B = \tau + \kappa + 2S$, and $\delta = A/B$,
then $K/p = A\cos(\phi - \theta) + B$
and $p = K/\{B[1 + \delta\cos(\phi - \theta)]\}$.

The instantaneous pressure p is

(a) a minimum, when $\phi = \theta$, i.e. $(\phi - \theta) = 0$,

(b) a maximum, when $\phi = (\theta + \pi)$, i.e. $(\phi - \theta) = \pi$,

therefore, $p_{min} = K/[B(1 + \delta)]$, and $p_{max} = K/[B(1 - \delta)]$.

Thus, $$p = p_{max}(1-\delta)/[1+\delta\cos(\phi-\theta)] \tag{4.10a}$$

$$= p_{min}(1+\delta)/[1+\delta\cos(\phi-\theta)], \tag{4.10b}$$

and the pressure ratio $p_r = p_{max}/p_{min} = (1+\delta)/(1-\delta)$. (4.11)

Mean cycle-pressure

The mean cycle-pressure is given by

$$p_{mean} = (1/2\pi)\int_0^{2\pi} p \, d(\phi-\theta)$$

$$= (1/2\pi)\int_0^{2\pi} \{p_{max}(1-\delta)/[1+\delta\cos(\phi-\theta)]\} \, d(\phi-\theta) \tag{4.12}$$

which can be resolved to

$$p_{mean} = p_{max}[(1-\delta)/(1+\delta)]^{\frac{1}{2}}. \tag{4.13}$$

Heat transferred and work done

Since the processes of expansion and compression take place isothermally the heat transferred Q is equal to the work done P, therefore

$$Q = P = \int p \, dV.$$

If $V = \frac{1}{2} V_E(1+\cos\phi)$,

$$dV = -\tfrac{1}{2} V_E \sin\phi \, d\phi \tag{4.14}$$

and, if

$$p = p_{mean}[1-\Delta\cos(\phi-\theta)], \text{ approximately,} \tag{4.15}$$

where $\Delta = 2\delta/[1+(1-\delta^2)^{\frac{1}{2}}]$,

then $$Q = -\tfrac{1}{2}\int_0^{2\pi} \{p_{mean} V_E[1-\Delta\cos(\phi-\theta)] \sin\phi\} \, d\phi$$

$$= -\tfrac{1}{2} p_{mean} V_E \int_0^{2\pi} [\sin\phi - \Delta(\cos\phi\cos\theta\sin\phi + \sin\theta\sin^2\phi)] \, d\phi$$

$$= -\tfrac{1}{2} p_{mean} V_E \left[-\cos\phi - \Delta[-\cos\theta \cdot \tfrac{1}{2}\cos 2\phi + \sin\theta(\tfrac{1}{2}\phi - \tfrac{1}{4}\sin 2\phi)]\right]_0^{2\pi}$$

$$= -\tfrac{1}{2} p_{mean} V_E \left[-\Delta\sin\theta \frac{\phi}{2}\right]_0^{2\pi}$$

$$= -\tfrac{1}{2}\pi p_{mean} V_E \Delta\sin\theta. \tag{4.16}$$

Expansion space

The variation in volume in the expansion space follows the equation

$$V_e = \tfrac{1}{2} V_E (1 + \cos \phi),$$

which conforms to the required equation (eqn (4.14)), therefore the heat transferred in the expansion space, from eqn (4.16), is given by

$$Q = \pi p_{\text{mean}} V_E \delta \sin \theta / [1 + (1 - \delta^2)^{\frac{1}{2}}]. \tag{4.17}$$

Compression space

The variation in volume of the compression space follows the equation

$$V_c = \tfrac{1}{2} \kappa V_E [1 + \cos (\phi - \alpha)] \tag{4.18}$$

and, by a process similar to that above, we can obtain expressions for the pressure and volume in the required form, so that heat transferred in the compression space is given by

$$Q_c = [\pi p_{\text{mean}} V_E \kappa \delta \sin (\theta - \alpha)] / [1 + (1 - \delta^2)^{\frac{1}{2}}]. \tag{4.19}$$

Dividing eqn (4.19) by eqn (4.17),

$$\begin{aligned} Q_c/Q &= [\kappa \sin (\theta - \alpha)] / \sin \theta \\ &= \kappa (\sin \theta \cos \alpha - \cos \theta \sin \alpha) / \sin \theta \\ &= \kappa (\cos \alpha - \sin \alpha / \tan \theta), \end{aligned}$$

but $\tan \theta = \kappa \sin \alpha / (\tau + \kappa \cos \alpha)$ and, therefore, $Q_c/Q = -\tau$.

The heat transferred in the expansion space is of opposite sign to the heat transferred in the compression space, and is numerically different by the temperature ratio τ. By analogy, the work done in the two spaces has the same relationship, $P_C = -\tau P_E$, and the net power is $P = P_E + P_C = (1 - \tau) Q$.

In the case of the machine acting as a prime mover $T_E > T_C$, i.e. $\tau < 1$, and the thermal efficiency η = (heat supplied − heat rejected)/(heat supplied)

$$= (Q - \tau Q)/Q = 1 - \tau = (T_E - T_C)/T_E.$$

This corresponds to the Carnot efficiency.

When the machine acts as a refrigerator, $T_C > T_E$, i.e. $\tau > 1$, and

$$\begin{aligned} \text{the coefficient of performance} &= \text{heat lifted/work done} \\ &= Q/(Q - Q_c) = 1/(1 - \tau) \\ &= T_E/(T_E - T_C). \end{aligned}$$

For a heat pump, $T_C > T_E$, i.e. $\tau > 1$, and the

$$\begin{aligned} \text{coefficient of performance} &= \text{heat rejected/work done} \\ &= Q_c/(Q - Q_c) = \tau/(1 - \tau) = T_C/(T_E - T_C). \end{aligned}$$

This corresponds to the inverse thermal efficiency.

Mass distribution in the machine

From the characteristic gas equation,

$$M = pV/RT,$$

where $p = p_{mean}(1 - \delta^2)^{\frac{1}{2}}/[1 + \delta \cos(\phi - \theta)]$.

(a) *Expansion space* $V_e = \frac{1}{2} V_E(1 + \cos\phi)$.

The instantaneous mass of working fluid in the expansion space is given by

$$M_e = \tfrac{1}{2} V_E p_{mean} (1 - \delta^2)^{\frac{1}{2}} (1 + \cos\phi)/[R\, T_E(1 + \delta \cos(\phi - \theta)]. \qquad (4.20)$$

The rate of change of mass of working fluid in the expansion space is

$$dM_e/d\phi = V_E p_{mean}(1 - \delta^2)^{\frac{1}{2}} \{\delta [\sin(\phi - \theta) - \sin\theta] - \sin\phi\}/2RT_E \, [1 + \delta \cos(\phi - \theta)]^2.$$

(b) *Compression space* $V_c = \frac{1}{2} \kappa V_E [1 + \cos(\phi - \alpha)]$.

The instantaneous mass of working fluid in the compression space is given by

$$M_c = \tfrac{1}{2} [\kappa V_E p_{mean} (1 - \delta^2)^{\frac{1}{2}} (1 + \cos(\phi - \alpha)]/2RT_C \, [1 + \delta \cos(\phi - \theta)]. \qquad (4.21)$$

The rate of change of mass of working fluid in the compression space is

$$dM_c/d\phi = \kappa V_E p_{mean}(1 - \delta^2)^{\frac{1}{2}} \{\delta [\sin(\phi - \theta) + \sin(\alpha - \theta) - \sin(\theta - \alpha)]\}/2RT_C [1 + \delta \cos(\phi - \theta)]^2.$$

(c) *Dead space* ($V_D = X V_E$, constant).

The instantaneous mass of the working fluid in the dead space is given by

$$M_d = [X V_E p_{mean}(1 - \delta^2)^{\frac{1}{2}}]/RT_D [1 + \delta \cos(\phi - \theta)]. \qquad (4.22)$$

The rate of change of mass of working fluid in the dead space is

$$dM_d/d\phi \; [X V_E p_{mean} (1 - \delta^2)^{\frac{1}{2}} \delta \sin(\phi - \theta)]/RT_D [1 + \delta \cos(\phi - \theta)]^2.$$

Now $dM_e + dM_c + dM_d = 0$, so that the total mass of working fluid M_T is constant. Now,

$$M_T = V_E p_{mean} (1 - \delta^2)^{\frac{1}{2}} \{\tau(1 + \cos\phi) + \kappa [1 + \cos(\phi - \alpha)] + 2S\}/2RT_C [1 + \delta \cos(\phi - \theta)]$$

and, when $\phi = 0$,

$$M_T = V_E P_{mean} (1 - \delta^2)^{\frac{1}{2}} [\tau + S + (\kappa/2)(1 + \cos\alpha)]/RT_C(1 + \delta \cos\theta). \qquad (4.23)$$

Heat lifted and engine output in dimensionless units

(a) The heat lifted per unit mass of working fluid, combining eqns (4.17) and (4.23) is given by

$$Q_{mass} = Q/RT_C = \pi\delta \sin\theta(1 + \cos\theta)/ \{(1-\delta^2)^{\frac{1}{2}} [1 + (1-\delta^2)^{\frac{1}{2}}] [\tau + (\kappa/2)(1 + \cos\alpha) + S]\}.$$

Similarly, the net engine-output per unit mass of working fluid is given by

$$P_{mass} = P/RT_C = (\tau - 1)Q_{mass}. \tag{4.25}$$

(b) Non-dimensional expressions, in terms of characteristic pressures and volumes, may be devised as follows. The combined swept volume is given by

$$V_T = (V_E + V_C) = (1 + \kappa)V_E.$$

Combining this with eqns (4.13) and (4.18), then

$$Q_{max} = Q/(P_{max}V_T) = [\pi(1-\delta)^{\frac{1}{2}}\,\delta \sin\theta] / [(1+\kappa)(1+\delta)^{\frac{1}{2}}\,(1 + 1 - \delta^2)]^{\frac{1}{2}} \tag{4.26}$$

and

$$P_{max} = (\tau - 1)Q_{max}. \tag{4.27}$$

Advanced theoretical analysis

Clearly, analyses as approximate as the Schmidt type are of limited value, since the predictions must be modified by a multiplier in the range 0·3 to 0·4, in order to establish realistic values. Considerable effort has been invested, therefore, in the development of improved analyses. A detailed discussion is beyond the scope of this work, but a brief review is given, principally to assist those wishing to specialize in this field.

Little is known of the Philips Company's work on theoretical development. Meijer (1969a) and Köhler (1955), writing about prime movers and cooling engines, have resorted to Schmidt analysis (or a derivative thereof) for illustration. Schalkwijk (1959) and Köhler (1965) have published material concerned with the operation of Stirling-cycle regenerative heat-exchangers. Meijer has referred to advanced theoretical developments and computer simulation of engines for design optimization: no details have yet been disclosed, nor are they likely to be revealed in the near future. The ability to perform a close simulation of an actual engine is, perhaps, one of the principal attractions for potential Philips licensees, and this technique may be expected to remain a close commercial secret.

Amongst the openly-published advanced theoretical analyses of the last twenty years, undoubtedly the most consistent effort has been made by Finkelstein; the work is reported in the series of publications listed in the bibliography.

Working from an established basis of Schmidt-type analyses, Finkelstein's first development was a generalized analysis which allowed for the introduction of non-isothermal processes of expansion and compression. The theory was explored by Walker and Khan (1965), with particular emphasis on the limiting case of adiabatic compression and expansion. For example, it was found that a prime mover, having a Schmidt–Carnot efficiency of 50 per cent with isothermal compression and expansion processes, had an efficiency of 34·3 per cent with adiabatic processes. Similarly, the coefficient of performance of a refrigerator was reduced from 1, with isothermal processes, to 0·543, with adiabatic processes. Furthermore, the thermal efficiency and coefficient of performance become functions of all the design parameters (including α, κ and X) although, in the Schmidt theory, it is a function of τ only. The potential of this simple theory was never completely exploited, and it is likely that much could be gained by its further development.

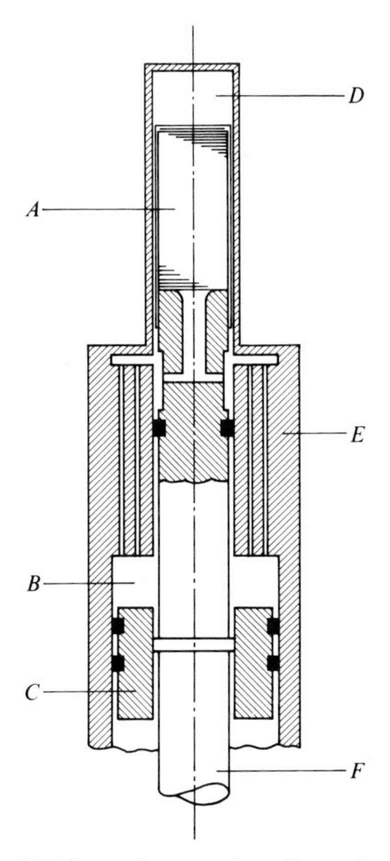

Fig. 4.1. Elements of a small Stirling-cycle cryogenic cooling engine.
A – regenerator, *B* – compression space, *C* – piston, *D* – expansion space, *E* – cooler, *F* – displacer rod.

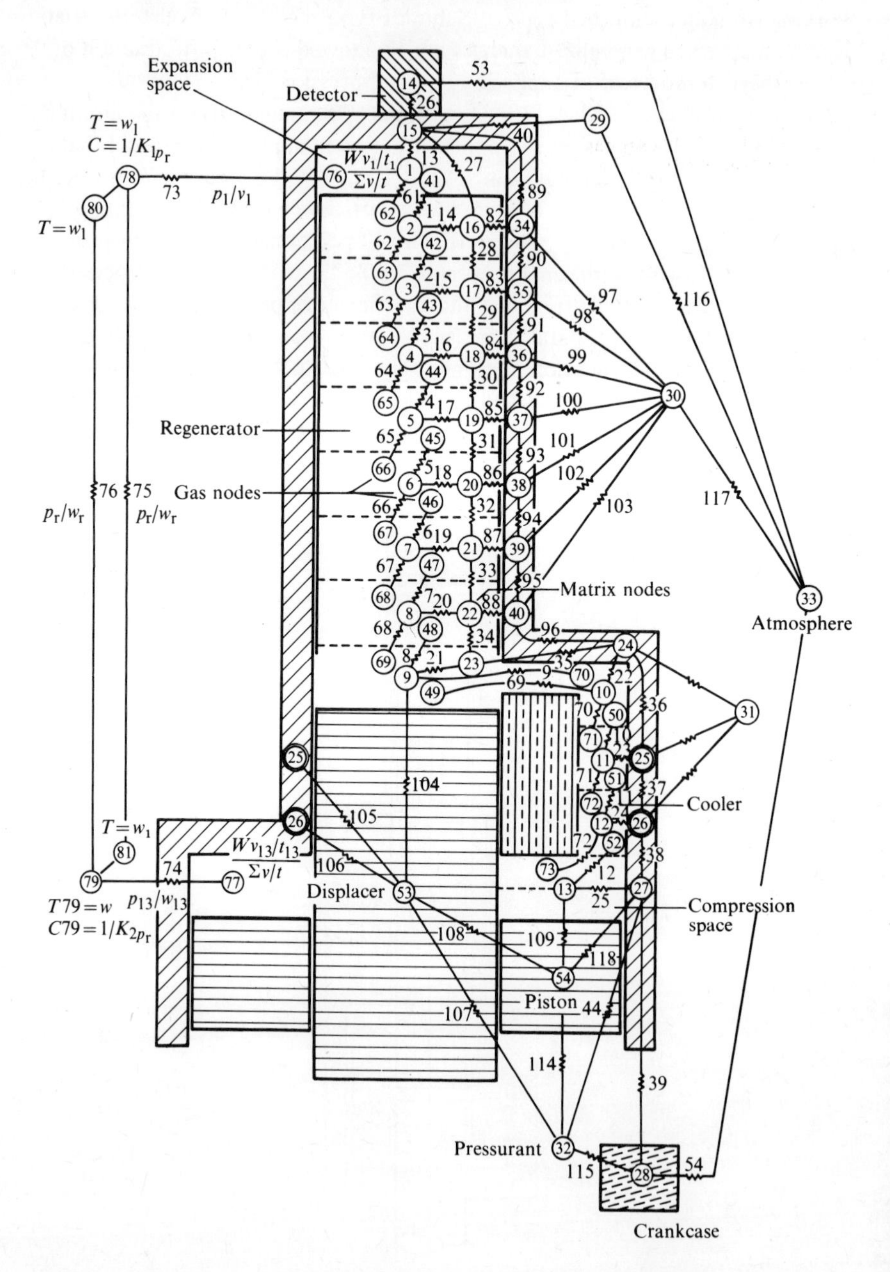

Fig. 4.2. Representation of the nodal network for digital computer simulation of a small Stirling-cycle cryogenic cooling engine. (After Finkelstein, Walker, and Joshi 1970.)

A further advance came with a publication by Finkelstein (1964), outlining a technique for giving a detailed mathematical description of a regenerative thermal machine in terms of 39 parameters and 4 functional relationships, where the solution is given in terms of 28 dependent variables. So far as is known, this theory has not been explored systematically.

Subsequently, Finkelstein became concerned with simulation by analogue computer and, more recently, with adaptation and modification of the well-known TAP (Thermal Analysis Program) to Stirling-cycle machines. This simulation program has been refined and made operational by Finkelstein, and has been used by the writer in the optimization of miniature cooling engines. A fairly detailed description of the program has been given by Finkelstein, Walker, and Joshi (1970). The program requires that a machine be designed and specified in considerable detail. It is then broken down into a series of nodal control volumes which are sufficiently small to permit accurate treatment of the temperature and pressure changes, yet not large enough to cause appreciable errors to arise from the substitution of a set of discrete points for continuous functions. The gas spaces, the machine elements, and the external environment are all treated as a fixed number of isothermal zones, of calculated thermal capacity, between which heat can flow, either by conduction through the solid or by surface convection as a result of the computed temperature-differences. For example, Fig. 4.1 shows the elements and configuration of a small cryogenic cooling engine, and Fig. 4.2 is a representation of the nodal network for simulation of a cryogenic cooling engine. The program is a remarkably powerful simulation of the operation of a regenerative engine, and could be adapted to any configuration of open-cycle or closed-cycle, machine. For operation, a high-speed large-storage digital computer is required; when used at the University of Calgary, each cycle required somewhere between 6 and 40 minutes of I.B.M. 360/50 computer-time. The principal difficulty is that much of the information on heat transfer and fluid flow, required for the data input, is not known. Thus, it is not possible to assess the accuracy of the simulation without experimental results, and when the experimental machine is available the need for computer-simulation largely disappears. Of course, it can be argued that, once the program is validated by comparison and adjustment of the predicted values to close agreement with the experimental values, it can be used to optimize the design of the experimental unit. This is partially true, but the order of uncertainty increases as conditions depart from those of the experimental machine, because the 'fudge' factors used to validate the model do not remain constant.

Other interesting contributions to advanced theoretical analyses have been made by Qvale and Smith (1969), and Rios and Smith (1969). They consider a basic cycle with adiabatic compression and expansion, and then separately assess the effects of irreversibilities. This approach allows for independent study of individual heat-exchange components in a series of successive approximations that can be extended to the required degree of complexity.

5 Preliminary engine design

Principal design parameters

The principal independently-chosen design parameters of a Stirling engine are:

(1) The temperature ratio $\tau = T_C/T_E$, the ratio of temperatures in the compression and expansion spaces.
(2) The swept-volume ratio $\kappa = V_C/V_E$, the ratio of swept volume in the compression and expansion space.
(3) The dead-volume ratio $X = V_D/V_E$, the total internal volume of heat-exchangers (and associated ducts and ports) expressed as multiples of the swept volume in the expansion space.
(4) The phase angle α by which volume variations in the expansion space *lead* those in the compression space.
(5) The pressure of the working fluid, expressed as the maximum or mean pressure.
(6) The speed of the engine N.
(7) The bore and stroke of the reciprocating member in the expansion space.

Summary of Schmidt-cycle design equations

(8) Instantaneous volume of expansion space $V_e = \frac{1}{2}V_E(1 + \cos\phi)$.
(9) Instantaneous volume of compression space $V_c = \frac{1}{2}\kappa V_E[1 + \cos(\phi - \alpha)]$.
(10) Instantaneous volume of total working space $= V_e + V_c + V_D$.
(11) Instantaneous pressure $p = p_{max}(1-\delta)/[1 + \delta\cos(\phi - \theta)]$,

$$\text{where } \delta = (\tau^2 + 2\tau\kappa\cos\alpha + \kappa^2)^{\frac{1}{2}}/(\tau + \kappa + 2S)$$
$$\tan\theta = \kappa\sin\alpha/(\tau + \cos\alpha)$$
$$S = 2X\tau/(\tau + 1).$$

(12) Pressure ratio $p_{max}/p_{min} = (1+\delta)/(1-\delta)$.
(13) Mean pressure $p_{mean} = p_{max}[(1-\delta)/(1+\delta)]^{\frac{1}{2}}$.

For a prime mover, $T_E > T_C$.

(14) Net cycle-power per cycle

$$P = (p_{max}V_T)\pi\frac{(\tau - 1)}{(\kappa + 1)}\left(\frac{1-\delta}{1+\delta}\right)^{\frac{1}{2}}\frac{\delta\sin\theta}{[1 + (1-\delta^2)^{\frac{1}{2}}]}$$

where $V_T = (V_E + V_C) = (1 + \kappa)V_E$.

(15) Power per unit mass of working fluid

$$P_{\text{mass}} = \frac{RT_{\text{C}}\,\pi(\tau - 1)(1 + \delta\cos\theta)(\delta\sin\theta)}{(1-\delta^2)^{\frac{1}{2}}\,[1 + (1-\delta^2)^{\frac{1}{2}}]\,[\tau + \frac{\kappa}{2}(1 + \cos\alpha) + S]}$$

and thermal efficiency

$$\eta = \frac{T_{\text{E}} - T_{\text{C}}}{T_{\text{E}}} = (1 - \tau).$$

(16) Heat transferred in expansion space, per cycle,

$$Q_{\text{E}} = \pi p_{\text{mean}}\,V_{\text{E}}\frac{\delta\sin\theta}{1 + (1-\delta^2)^{\frac{1}{2}}}.$$

(17) Heat transferred in the compression space, per cycle,

$$Q_{\text{C}} = -\tau Q_{\text{E}}.$$

For a refrigerating machine, $T_{\text{E}} < T_{\text{C}}$.

(18) Heat lifted in cold expansion space, per cycle,

$$Q_{\text{E}} = (p_{\max}V_{\text{T}})\frac{\kappa}{(\kappa + 1)}\left(\frac{1-\delta}{1+\delta}\right)^{\frac{1}{2}}\frac{\delta\sin\theta}{[1 + (1-\delta^2)^{\frac{1}{2}}]}.$$

(19) Heat lifted per unit mass of working fluid

$$Q_{\text{mass}} = \frac{RT_{\text{C}}\,\pi(1 + \delta\cos\theta)\,\delta\sin\theta}{(1-\delta^2)^{\frac{1}{2}}\,(1 + (1-\delta^2)^{\frac{1}{2}})\,[\tau + \frac{\kappa}{2}(1 + \cos\alpha) + S]}$$

(20) Coefficient of performance (COP)

$$= \frac{T_{\text{E}}}{(T_{\text{E}} - T_{\text{C}})} = \frac{1}{(1-\tau)}.$$

(21) Heat transferred from the compression space (to cooling medium)

$$Q_{\text{C}} = \tau Q_{\text{E}}.$$

(22) Power required to drive the refrigerator $P = (1 - \tau)Q_{\text{E}}$.

For a heat pump, $T_{\text{E}} < T_{\text{C}}$.

(23) Heat transferred from the hot (compression) space per cycle,

$$Q_{\text{C}} = (p_{\max}V_{\text{T}})\frac{\pi\tau}{(\kappa + 1)}\left(\frac{1-\delta}{1+\delta}\right)^{\frac{1}{2}}\frac{\delta\sin\theta}{[1 + (1-\delta^2)^{\frac{1}{2}}]}$$

(24) Heat transferred from the hot (compression) space, per unit mass of working fluid,

$$Q_{\text{mass}} = \frac{RT_c \pi \tau (1 + \delta \cos \theta)\, \delta \sin \theta}{(1 - \delta^2)^{\frac{1}{2}} \left(1 + (1 - \delta^2)^{\frac{1}{2}}\right) \left(\tau + \frac{\kappa}{2}(1 + \cos \alpha) + S\right)}$$

(25) Coefficient of performance (COP) $= T_C/(T_E - T_C) = \tau/(1 - \tau)$.

(26) Heat transferred from the expansion space (heat source),

$$Q_E = Q_C/\tau.$$

(27) Power required to drive the heat pump,

$$P = Q_C(1 - \tau)/\tau.$$

Optimization of design parameters

It is obvious from the Schmidt-cycle equations (summarized above) that the net cycle-power and the thermal loads on the heat-exchangers are direct linear functions of engine *speed* (N), *pressure* of the working fluid (p_{max}), and *size* of

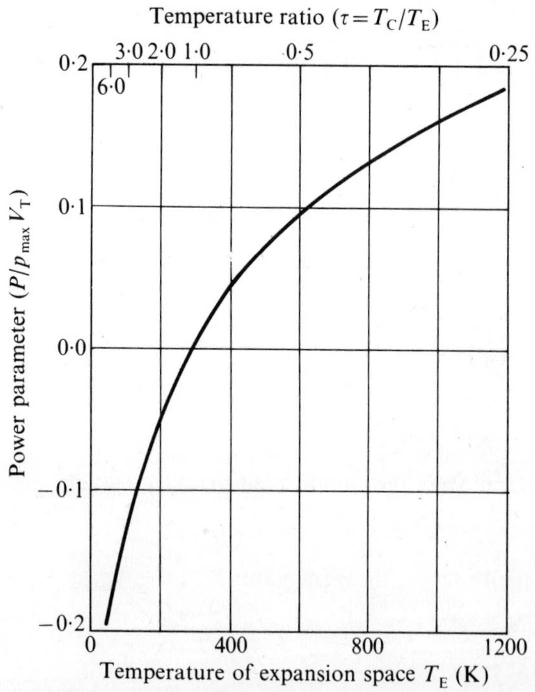

Fig. 5.1. Effect of temperature on cycle power. The figure shows the effect on the non-dimensional power parameter $P/(p_{\text{max}} V_T)$ of variation in the expansion space temperature T_E, with T_c constant at 300K, phase angle $\alpha = 90°$, swept-volume ratio $\kappa = 0{\cdot}8$ and dead-volume ratio $X = 1{\cdot}0$. At expansion-space temperatures below 300 K, $\tau > 1{\cdot}0$ and the power parameter is negative, because the cycle is acting as a cooling cycle, requiring a net input of work.

the engine, expressed in terms of the total swept volume V_T. The effect which the four principal design parameters (τ, κ, α and X) have on performance is less obvious. In particular, it is not clear what combination of these should be used to achieve optimum performance. This is an important consideration, since these parameters must be determined at the design stage and, except for the temperature ratio τ, cannot readily be varied afterwards, except by structural changes in the machine.

Figs. 5.1–5.4 show the effect on the cycle-power parameter $P/p_{max} V_T$ of independent variation of one of the four parameters τ, κ, α and X, with the other three maintained constant. In Fig. 5.1 the effect of the temperature-ratio on cycle-power is explored for expansion-space temperatures (T_E) both above and below the assumed compression-space temperature of 300K, thus embracing both refrigerating machines and prime movers. With $T_E > T_C$, the power parameter is positive, and progressively increases as the expansion-space temperature increases. When $T_E < T_C$, the machine is acting as a refrigerator and, as the expansion-space temperature decreases, there is a progressive increase in the power required to drive the machine. It is clear that for *engines* the available power may be increased by the use of high-temperature materials for the expansion-space cylinder and heat-exchanger, and for *refrigerators* the temperature of refrigeration should be as high as possible. Fig. 5.2. shows the effect of the swept-volume ratio κ on the power parameter. The curves show clearly that, for

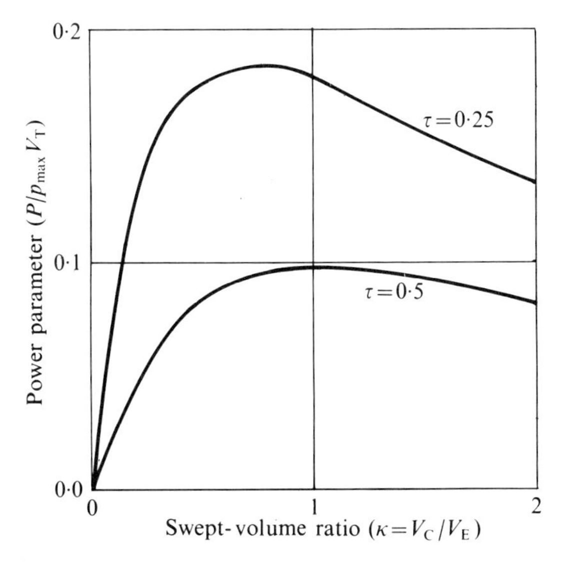

Fig. 5.2. Effect of the swept-volume ratio κ on cycle power. The figure shows the effect on the non-dimensional power parameter $P/(p_{max} V_T)$ of variation in the swept-volume ratio κ, with constant values of the temperature τ = 0·25 and 0·5, phase angle α = 90°, and dead-volume ratio X = 1·0. On the two curves shown, for different values of τ, the maxima do not occur at the same value of κ. There is no single 'best' value for κ since the optimum value depends on the combinations α, τ, and X.

given values of τ, α, and X, there is a definite optimum value of κ at which the power parameter is a maximum. Comparison of the two curves for $\tau = 0{\cdot}25$ and $0{\cdot}5$ shows, however, that the optimum value of κ is not constant, but changes from about $0{\cdot}75$, when $\tau = 0{\cdot}25$, to about $1{\cdot}0$, when $\tau = 0{\cdot}5$. Changes in α and X also produce an adjustment of the optimum value of κ. Thus, there is no single 'best' value of κ. Fig. 5.3 shows the effect of the dead-volume ratio $X = V_D/V_E$

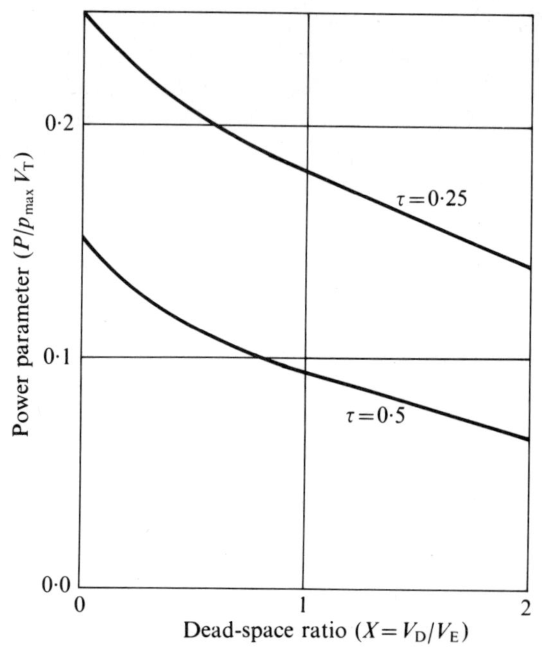

Fig. 5.3. Effect of the dead-space ratio X on cycle power. The figure shows the effect on the non-dimensional power parameter $P/(p_{max} V_T)$ of variation in the dead-space ratio X, with constant values of τ, κ, and α. The dead space is the porous volume of the regenerator and other heat-exchangers, connecting ducts, and the clearance volumes in the expansion and compression spaces. Increase in the dead volume decreases the ratio of maximum volume to minimum volume which decreases the range of the pressure excursion, thus causing a decrease in the cycle power. The dead space *must* be minimized for high cycle-power.

on the power parameter. The message of this figure is very clear: increase in the dead space above the absolute minimum required reduces the power parameter. The dead space must be kept as small as possible. Fig. 5.4 shows the effect of the phase angle α on cycle power. The power parameter is remarkably insensitive to changes in the phase angle over an extended range from 60° to 120° of crank rotation. For the particular assumed conditions the optimum value was between 90° and 115°.

A three-dimensional presentation is given in Fig. 5.5 of the variations in engine-power parameter with change in both phase angle α and swept-volume ratio κ, for constant values of τ and X. This presentation results in the generation of a solid surface. Any changes in τ or X cause the generation of a series of similar, but different, overlaying surfaces. The apex of the surface represents the maximum value of the power parameter, and occurs at the optimum combination of swept-volume ratio and phase angle, for the given values of τ and X. Fig. 5.5 shows two surfaces, generated with the different power parameters $P/p_{max}V_T$ and P/RT_C. For these two surfaces, the apex occurs at different combinations of phase angle and swept-volume ratio. In the case of the surface drawn for the power parameter $P_{mass} = P/RT_C$, $\alpha_{opt} = 0{\cdot}45\ \pi$ rad and $\kappa_{opt} = 2{\cdot}9$. In the case of the surface drawn for the power parameter $P_{max} = P/p_{max}V_T$, $\alpha_{opt} = 0{\cdot}54\ \pi$rad

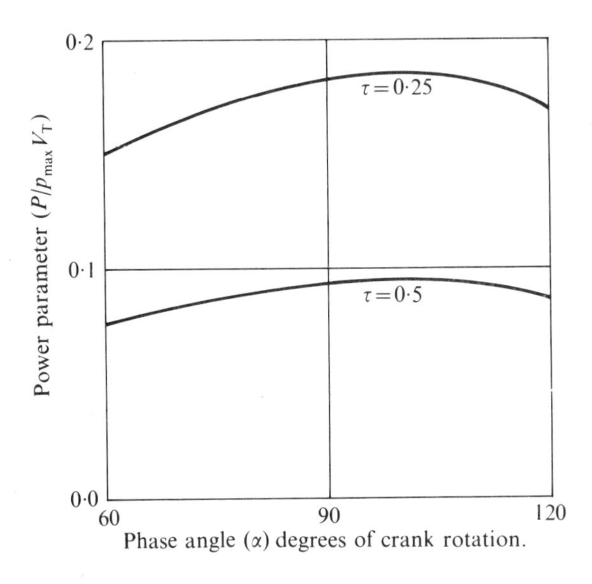

Fig. 5.4. Effect of the phase angle α on cycle power. The figure shows the effect on the non-dimensional power parameter $P/(p_{max}V_T)$ of variation in the phase angle α, with constant values of τ, κ and X. The power parameter is remarkably insensitive to variation in α, this permits considerable flexibility in the geometrical design of Stirling-engine drive mechanisms.

and $\kappa_{opt} = 0{\cdot}74$. A simple explanation for this strange phenomenon is that optimization of design on the basis of the power parameter P/RT_C results in a machine configuration that makes the best possible use of a *limited mass of working fluid.* Optimization of design on the basis of parameter $P/p_{max}V_T$ results in a machine configuration of the maximum possible power within limits of the *maximum pressure and combined swept-volume.* The maximum pressure of the working fluid is an important design criterion, since this affects the strength, and hence the *weight,* of the machine structure, whereas the combined swept-volume V_T is indicative of the *size* of the structure. Clearly then, the power parameter $P/p_{max}V_T$ should be used for optimization purposes.

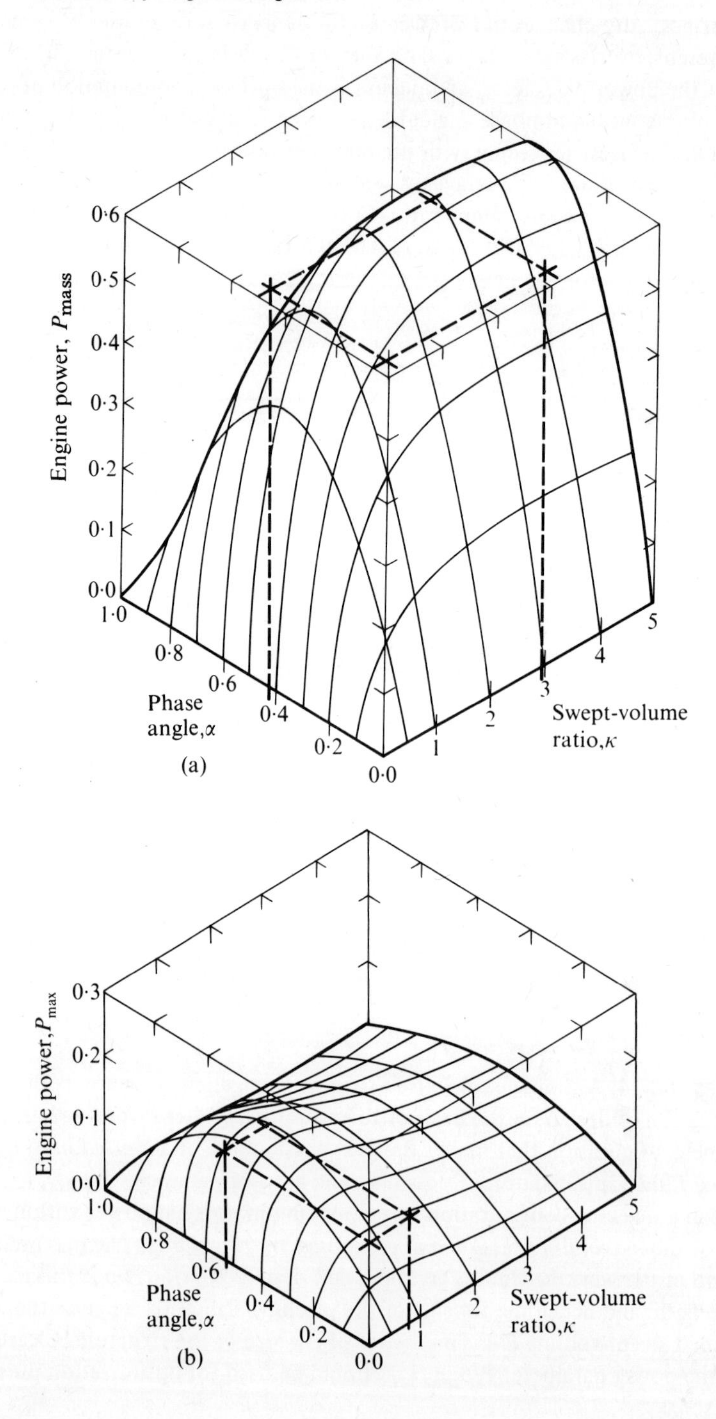
0·6
0·5
0·4
0·3
0·2
0·1
0·0
1·0
0·8
0·6
0·4
0·2
0·0
Engine power, P_{mass}
Phase angle, α
1
2
3
4
5
Swept-volume ratio, κ
(a)
0·3
0·2
0·1
0·0
1·0
0·8
0·6
0·4
0·2
0·0
Engine power, P_{max}
Phase angle, α
1
2
3
4
5
Swept-volume ratio, κ
(b)

However, once the basic machine configuration has been determined in terms of α, κ, τ and X, the power parameter P/RT_C may be used thereafter. Both it and $P/p_{max}V_T$ are equally applicable, and will return the same numerical values for the engine power per cycle P.

By way of example, Fig. 5.6 is a comparison of the work diagram obtained for the optimum combination of phase angle α and swept-volume ratio κ, with constant values of τ and X. In all three cases, the pressures are adjusted to be fractions of the maximum cycle-pressure, and so are comparable. Similarly, in all three cases, the maximum value of the total enclosed volume has been made identical at an arbitrary value of 4·6. In each case, the diagram at the extreme left is the work diagram for the expansion space, the one in the middle is for the compression space, and the diagram at the right is the total enclosed volume. Fig. 5.6a is the diagram resulting from the combination $\alpha = 0{\cdot}45\pi$ rad, $\kappa = 2{\cdot}9$ $\tau = 0{\cdot}3$, and $X = 1{\cdot}0$, the optimum configuration based on P/RT_C. Fig. 5.6(b) is the diagram resulting from a combination $\alpha = 0{\cdot}54\pi$ rad, $\kappa = 0{\cdot}74$, $\tau = 0{\cdot}3$, and $X = 1{\cdot}0$, the optimum combination based on $P/p_{max}V_T$. Fig. 5.6(c) is the diagram resulting from a modification to the configuration of Fig. 5.6(b) to adjust the absolute value of the dead-space volume V_D to be the same as in the configuration of Fig. 5.6(a).

The two machines represented in Figs. 5.6(a) and 5.6(b) are comparable therefore in terms of size and weight. The maximum pressure and the maximum total enclosed volume are the same, although the dead volume in one machine is nearly twice as great as in the other. Despite this, the net work-output of the machine in Fig. 5.6(b) is 1·38 times that of the machine in Fig. 5.6(a). When the dead volume of Fig. 5.6(b) is reduced to an amount comparable with that of Fig. 5.6(a) (as in machine represented by Fig. 5.6(c)) the ratio of net work-output increases to 2·24 in favour of the machine in Fig. 5.6(c). This example is a convincing justification for optimization on the basis of the power parameter $P/p_{max}V_T$.

Walker (1962) has drawn a corresponding comparison for refrigerating machines. Use of the cooling parameter $Q_E/p_{max}V_T$ for optimization studies is preferred, because it leads to a machine configuration having the maximum cooling capacity for a given size and weight.

Fig. 5.5. Effect of swept-volume ratio κ and phase angle α on cycle power. The figures are a three-dimensional representation of non-dimensional power parameters as a function of the swept-volume ratio κ and the phase angle α, with constant values of the temperature ratio $\tau = 0{\cdot}3$, and the dead-space ratio $X = 1{\cdot}0$. Fig. 5.5(a) is a representation for the power parameter P/RT_C (power per unit mass of working fluid) and Fig. 5.5(b) is a representation for the power parameter $P/(p_{max}V_T)$ (power for limited size and weight). The apex of the surfaces (the maximum value of the power parameter) occurs at the *optimum* values of phase angle α and swept-volume ratio κ.
In case (a) $\alpha_{opt} = 0{\cdot}45\ \pi$ radians and $\kappa_{opt} = 2{\cdot}9$.
In case (b) $\alpha_{opt} = 0{\cdot}54\ \pi$ radians and $\kappa_{opt} = 0{\cdot}74$.
In both (a) and (b) $\tau = 0{\cdot}3$ and $X = 1{\cdot}0$. (See Fig. 5.6.)

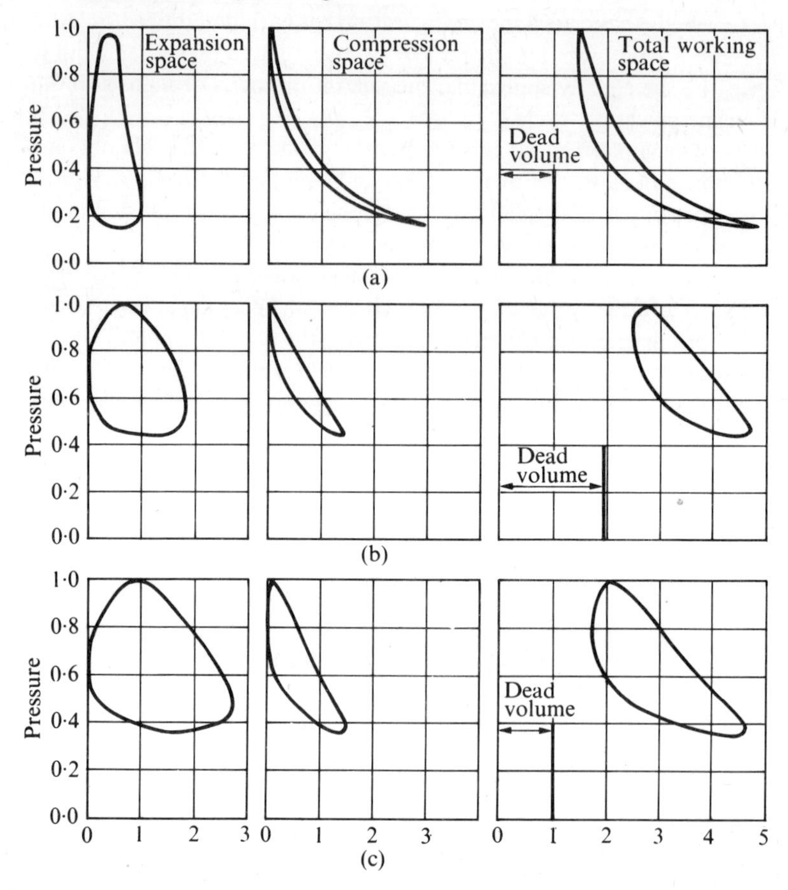

Fig. 5.6. Work diagrams for engines having the optimum combination of design parameters. The figure shows work diagrams for engines having optimum combinations of the design parameters α, λ. τ, and X, as determined by reference to the three-dimensional representation shown in Fig. 5.5. In all cases, the diagram at left is for the expansion space, the centre diagram is for the compression space, and the diagram at right is for the total working-space. Fig. 5.6(a) shows work diagrams for the cycle, optimized using the power parameter P/RT_c, with $\alpha_{opt} = 0{\cdot}45\ \pi$ radians, $\kappa_{opt} = 2{\cdot}9$, $\tau = 0{\cdot}3$, and $X = 1{\cdot}0$, as determined in Fig. 5.5(a). Fig. 5.6(b) shows work diagrams for the cycle, optimized using the power parameter $P/(p_{max} V_T)$, with $\alpha_{opt} = 0{\cdot}54\ \pi$ radians, $\kappa_{opt} = 0{\cdot}74$, $\tau = 0{\cdot}3$, and $X = 1{\cdot}0$ as determined in Fig. 5.5(b).† Fig. 5.6(c) shows work diagrams for a cycle having a combination similar to Fig. 5.6(b), but with the dead space reduced to the same over-all value as in Fig. 5.6(a) (same size of regenerator and heat-exchangers). The net cycle-output of this case is superior to case (a), by an even greater margin than case (b).

† For purposes of comparison, all the work diagrams are drawn with pressures as fractions of the maximum pressure, and with the same maximum value of the total working-space. Clearly, the area of diagram for the total working-space (the net cycle-work) is greater in case (b) than in case (a).

Consolidated design charts

Despite the interest and attraction of diagrams such as Fig. 5.5, it can readily be understood that there exists a virtually infinite array of possible permutations of engine design-parameters. It would be a tedious matter to search through the variations for an optimum combination. To overcome the difficulty, consolidated design charts have been prepared, and are presented in Fig. 5.7 (for prime movers) and Fig. 5.8 (for refrigerating machines).

Fig. 5.7 was prepared using the power parameter $P/p_{\text{max}} V_{\text{T}}$ as the basis for optimization. Surfaces, similar to that shown in Fig. 5.5, were generated for the value of $P/p_{\text{max}} V_{\text{T}}$, with different values of α and κ and constant values of τ and X. The apex of the surface was established, and the apex values of $P/p_{\text{max}} V_{\text{T}}$, α_{opt}, and κ_{opt} were plotted on appropriate charts, constituting Fig. 5.7. These were all drawn on the common basis of expansion-space temperature T_{E}, with the compression-space temperature *always* maintained constant at 300K. The apexes of other surfaces, with different values of τ and X, were determined, and plotted to obtain the complete consolidated chart. A similar technique was used to obtain the consolidated chart for the refrigerating machines, with optimization based on the cooling parameter $Q_{\text{E}}/p_{\text{max}} V_{\text{T}}$ (Fig. 5.8).

The work was carried out, using a self-optimizing digital computer program (with automatic use of recognized hill-climbing techniques), to locate the apex of the surface from any given fixed values of τ and X and starting values of α and κ, as described by Walker (1962).

Use of the consolidated chart for design

The design charts of Figs. 5.7 and 5.8 are recommended for use in the preliminary design stages of a Stirling-cycle machine.

In the case of a prime mover, it is necessary first to establish the permissible temperature in the expansion space T_{E}. This is dictated by the nature of the thermal source, and by the materials to be used for the expansion-space heat-exchanger and expansion cylinder. At the chosen temperature, a vertical line is drawn through the three charts. Where this line intersects the sets of curves, and at an appropriate value of the dead-space ratio X, the corresponding optimum value of both phase angle α and swept-volume ratio κ may be determined from the scales. The value of the engine-power parameter $(P/p_{\text{max}} V_{\text{T}})$ may be determined also. With T_{E} known, T_{C} may be estimated (it is about 300K with a water-cooled engine), so that τ can be calculated. With τ, X, α and κ thus established it is possible to proceed to the detailed design of the engine, utilizing the summary of design equations given earlier.

It cannot be over-emphasized that the predictions of Schmidt-cycle calculations are highly optimistic. Experience suggests that it is unwise to expect from a practical engine more than 30 to 40 per cent of the power and efficiency predicted by Schmidt-type analyses.

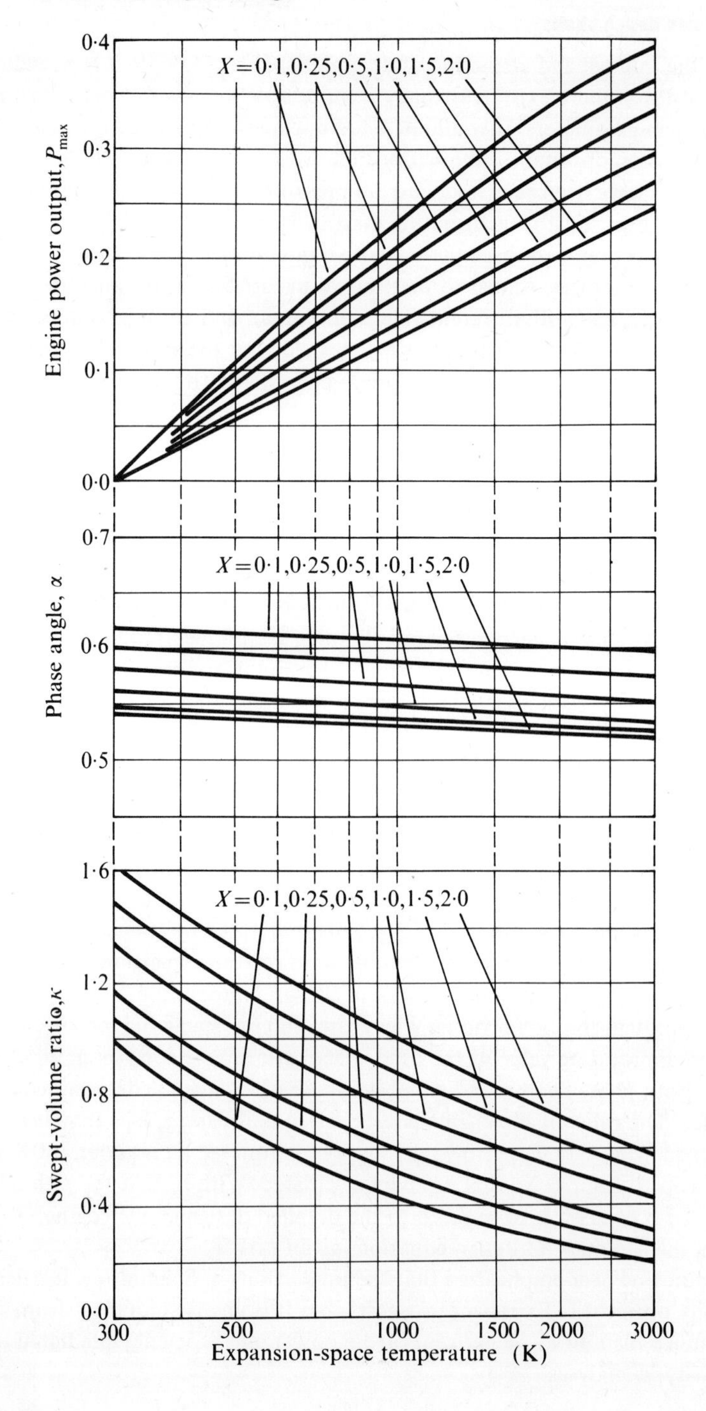

Engine power output, P_{max}
0·0
0·1
0·2
0·3
0·4
X=0·1, 0·25, 0·5, 1·0, 1·5, 2·0
Phase angle, α
0·5
0·6
0·7
X=0·1, 0·25, 0·5, 1·0, 1·5, 2·0
Swept-volume ratio, κ
0·0
0·4
0·8
1·2
1·6
X=0·1, 0·25, 0·5, 1·0, 1·5, 2·0
300
500
1000
1500
2000
3000
Expansion-space temperature (K)

Working fluid

In the Schmidt theory no explicit account is taken of the physical characteristics of the working fluid, except its behaviour as a perfect gas (i.e. it obeys the characteristic gas equation $PV = RT$). However, the assumptions on which the theory is based imply the use of an idealized working-fluid, having properties not found in nature. The assumption that there are no aerodynamic-friction losses could only be true if the working fluid were to have zero viscosity. Similarly, the assumptions of perfect regeneration and isothermal compression and expansion can only be attained if the working fluid were to have unreal values for specific heat and thermal conductivity.

In practice there appear to be only three working-fluids of significant interest: air, helium, and hydrogen. Air is of interest because it is so freely available. Helium and hydrogen are of interest because their thermophysical characteristics are such as to permit high rates of heat transfer and flow to occur, with relatively low aerodynamic-flow losses. In terms of engine performance, hydrogen is better than helium, and is also very much cheaper, but is highly combustible in the presence of air or oxygen.

Engines of high specific output and high thermal efficiency, operating at high pressures and high speeds (i.e. greater than 2000 rev./min), must use hydrogen or helium as the working fluid, in order to achieve the rates of heat- and mass-transfer necessary, with tolerable flow-losses. The sealing problems are very severe, however. Furthermore, the control systems needed to vary engine output are complicated, since they must incorporate reservoirs, valves, and, perhaps, a compressor to vary the pressure level, while conserving the working fluid. The cost of machines of this type are high, and applications are likely to be limited to relatively large engines, where the advantages of low noise (and pollution) levels justify increased cost, compared with internal-combustion engines. Cooling engines of high output (or those intended for refrigeration at cryogenic temperatures) must also use helium or hydrogen as the working fluid.

Engines using air as working fluid cannot achieve the high rates of heat- and mass-transfer found in hydrogen or helium engines. Such machines are, typically, large heavy engines of low specific output and low thermal efficiency. However, the working fluid can readily be replenished from atmospheric air, so that the sealing and materials problems are substantially eliminated, and the machines can be simple, cheap, and reliable. Air engines have such a poor performance

---

Fig. 5.7. Design chart for Stirling engines. The optimum combination of swept-volume ratio κ and phase angle α (in π radians) may be determined for given values of the dead-space ratio X and the temperature ratio τ. Assuming $T_c = 300\,K$, and with the metallurgical limit controlling the expansion space temperature T_E, draw a vertical line through T_E. Intersection of this line with the selected value for X occurs at the optimum values for κ and α. The appropriate value of the non-dimensional power parameter $P/p_{max}V_T$ is determined from the upper diagram. Practical engines may produce 0·3 to 0·4 of this predicted value.

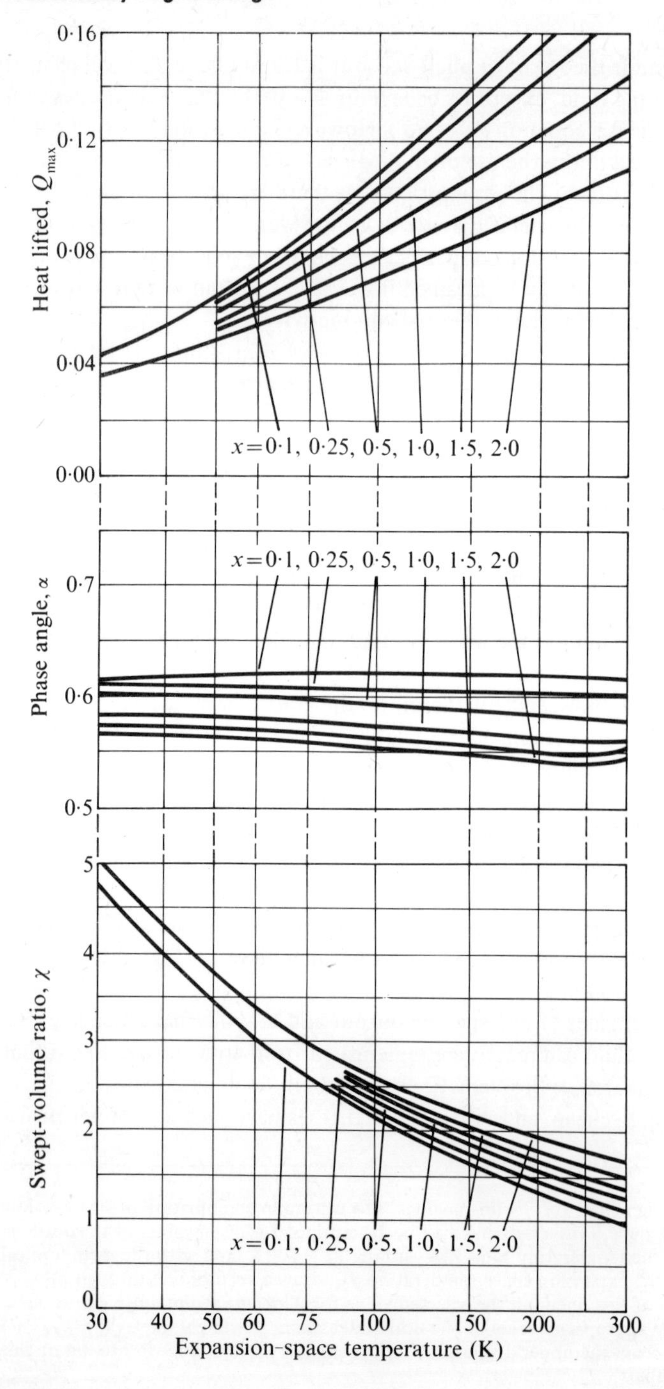
0·16
0·12
0·08
0·04
0·00
Heat lifted, Q_{max}
x=0·1, 0·25, 0·5, 1·0, 1·5, 2·0
0·7
0·6
0·5
Phase angle, α
x=0·1, 0·25, 0·5, 1·0, 1·5, 2·0
5
4
3
2
1
0
Swept-volume ratio, χ
x=0·1, 0·25, 0·5, 1·0, 1·5, 2·0
30 40 50 60 75 100 150 200 300
Expansion-space temperature (K)

that they offer no serious competition to internal-combustion engines, in either automotive or general-purpose application. There is, however, an urgent and increasing need for low-power (less than one horsepower) engines of high reliability and moderate efficiency, capable of operating unattended for long periods (in excess of one year) and utilizing fossil, or radioisotope, fuels. The engines are required to drive electric-power generators for navigational, meteorological, and telecommunications purposes. Stirling-cycle air engines appear admirably suited for this purpose.

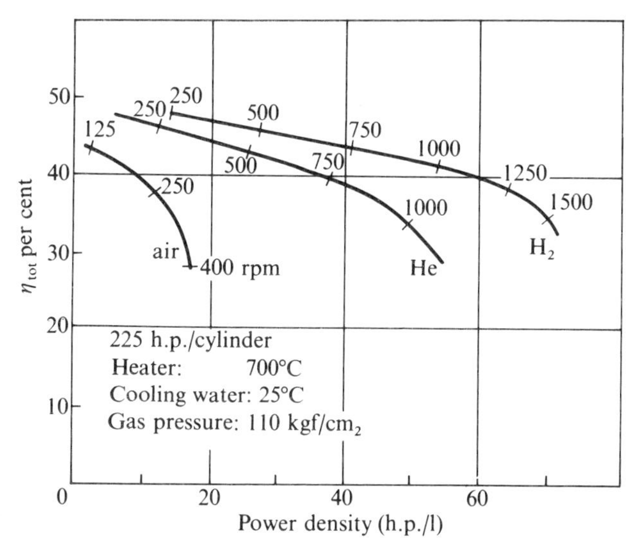

Fig. 5.9. Comparative performance of Stirling engines with air, hydrogen, and helium as the working fluid (after Meijer 1970). The figure shows that, at high speeds and high power-densities, hydrogen is the preferred working fluid, with helium the second choice. At low powers and low speeds, air may be used in small engines, with little loss of performance and appreciable practical advantages in terms of fluid sealing, replenishment, and simple engine design.

The comparative performance of Stirling engines with air, hydrogen, and helium is shown in Fig. 5.9. This is a reproduction of material presented by Meijer (1970), based on advanced simulation calculations for a single-cylinder Stirling engine of 225 h.p. The figure shows how the engine's thermal efficiency is related to power density (in terms of horsepower per litre of cylinder swept-volume) at different speeds and with three different working fluids (air, hydrogen, and helium). At high power-densities and high speeds, hydrogen is appreciably

---

Fig. 5.8. Design chart for Stirling-cycle cooling engines. Optimum values for swept-volume ratio κ and phase angle α may be determined for given values of the dead-space ratio X and the expansion space temperature T_E, with $T_c = 300\,K$ by drawing a vertical line on the chart through T_E. The cooling capacity, in terms of heat lifted in the expansion space, $C/p_{max}V_T$ will be optimistic by a factor of about three.

better than helium, and the curve for air is not able to reach the area of the diagram. However, the important point to note is that, at *low speeds* and *low power-densities,* there is no really significant difference between air, helium, and hydrogen. The selection of the working fluid is an important decision which has to be taken at the design stage, and requires the intended application of the engine to be clearly defined. If the intention is to make a high-speed high-performance engine then hydrogen or helium must be used, but if modest speed and performance can be tolerated then the use of air as the working fluid has considerable attractions, on grounds of simplicity and cost.

It is possible that other working fluids may be used in the future. Of interest, at present, are reacting fluids and two-phase two-component fluids; the latter is discussed briefly in Chapter 10.

6 Mechanical arrangements

Introduction

The elements of a Stirling engine include two spaces, at different temperature-levels, having volumes that can be varied cyclically; they are coupled through a regenerative heat-exchanger and auxiliary heat-exchangers. These simple elements can be combined in a bewildering range of mechanical arrangements, some of which have been identified (Finkelstein 1959) by the name of the inventor or original user. One key identifying feature is the manner in which the flow of the working fluid is controlled. There are two possibilities; flow is controlled either by valves or by volume changes. In some respects, the two types of machine are similar, but, in details of construction, operation, and fields of application, they are quite different.

In this present work, the name Stirling engine is limited to regenerative engines, where the flow is controlled by volume changes. Machines where the flow is controlled by valves are called Ericsson engines. These names are chosen somewhat arbitrarily in an attempt to bring some order to a confused situation, since no particular distinction has been established in the general literature. The names Stirling and Ericsson are themselves not entirely satisfactory for they suggest operation on ideal cycles, having isothermal compression and expansion and with regenerative transfer-processes, that are either constant pressure or constant volume. So far as is known, all the engines devised by Robert Stirling were of the closed-cycle type, where the flow is controlled by volume changes, whereas John Ericsson produced both types of machine.

Design variants of Stirling engines

All existing arrangements for the Stirling engine may be classified into two broad groups: (1) two-piston machines, and (2) piston–displacer machines. A further subdivision can be made in the latter group between machines in which the piston and displacer operate in a single cylinder, and those in which separate cylinders are provided. An example of each of these three arrangements is shown in Fig. 6.1. The principal distinction between a piston and a displacer is that pistons are (and displacers are not), provided with a (theoretically) gas-tight fluid seal, to prevent the passage of gas from one side to the other. Thus, the pressure of fluid, both above and below the displacer, is the same apart from aerodynamic-flow losses, and, when reciprocating, the displacer does no work on the gas, but merely displaces it from one side of the displacer to the other.

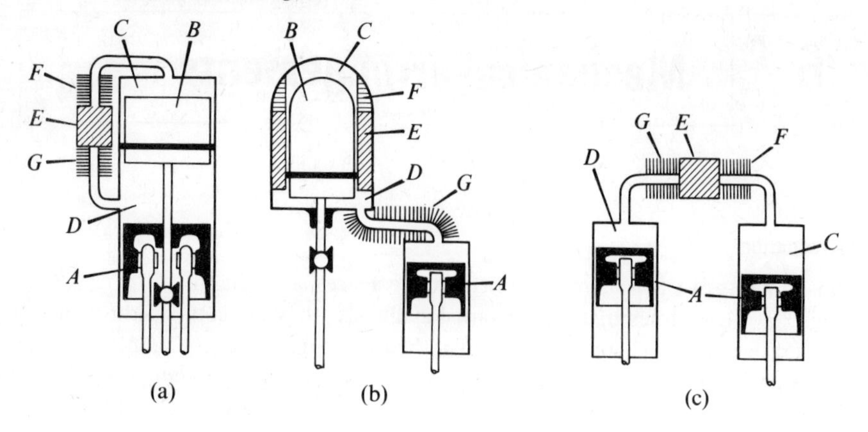

Fig. 6.1. Three basic arrangements by which most types of Stirling engine may be calculated.
(a) Piston–displacer in the same cylinder.
(b) Piston–displacer in separate cylinders.
(c) Two-piston machine.

A – piston, *B* – displacer, *C* – expansion space, *D* – compression space, *E* – regenerator, *F* – heater, *G* – cooler.

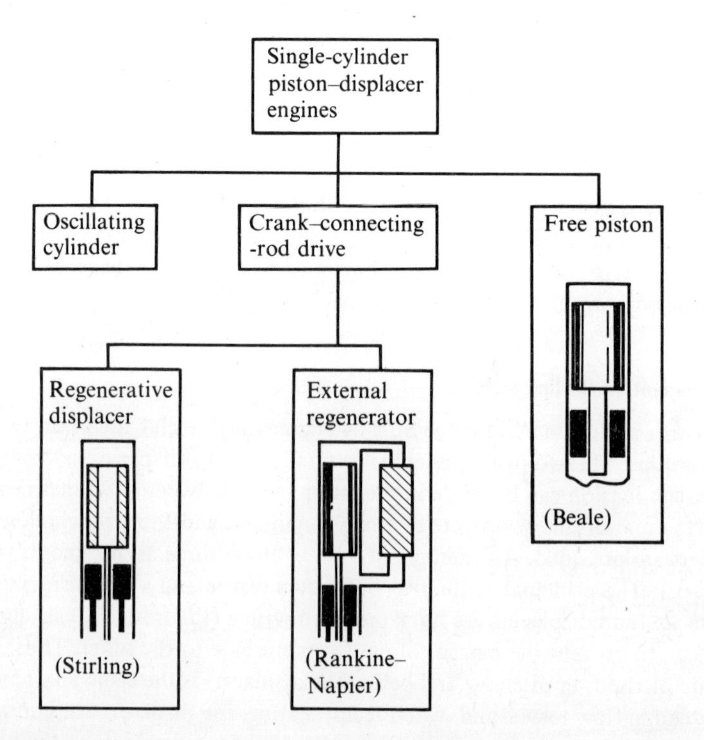

Fig. 6.2. Alternative arrangements of the single-cylinder piston–displacer engine.

In the case of a piston, the pressure of fluid, above and below the displacer, are *not* the same, except perhaps momentarily at some point in the cycle. Work is done on the gas by the piston, or on the piston by the gas, as the piston moves in the cylinder.

In some machines, the displacer is made up (either partly or wholly) by a porous metallic matrix, which itself constitutes the regenerative heat-exchanger, here called a regenerative displacer.

Single-cylinder piston–displacer machines

Some of the possible alternative arrangements of single-cylinder piston–displacer machines are shown in Fig. 6.2. This is a particularly favourable configuration; it was first used by Robert Stirling, in 1816, for the engine shown in Fig. 6.3. It has been used, also, for most of the machines developed by Philips, both prime movers and cooling engines.

Crank-driven engines can be of the type used by Stirling, with a regenerative displacer, or may have a separate external regenerator of the Rankine–Napier type. The possibility exists for the necessary volume variations to be gained by an oscillating-cylinder mechanism, but, so far as is known, machines of this type

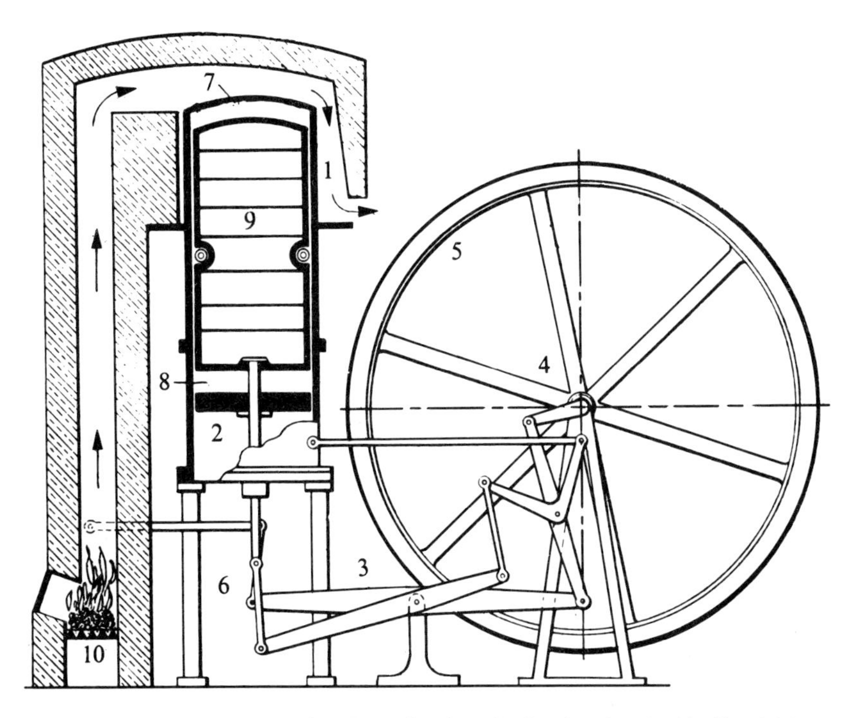

Fig. 6.3. The original Stirling engine. Reproduction of a drawing showing the first Stirling engine, from the original patent specifications of 1816. Such an engine was used in 1818 for pumping water from a quarry. (After Finkelstein 1959.)

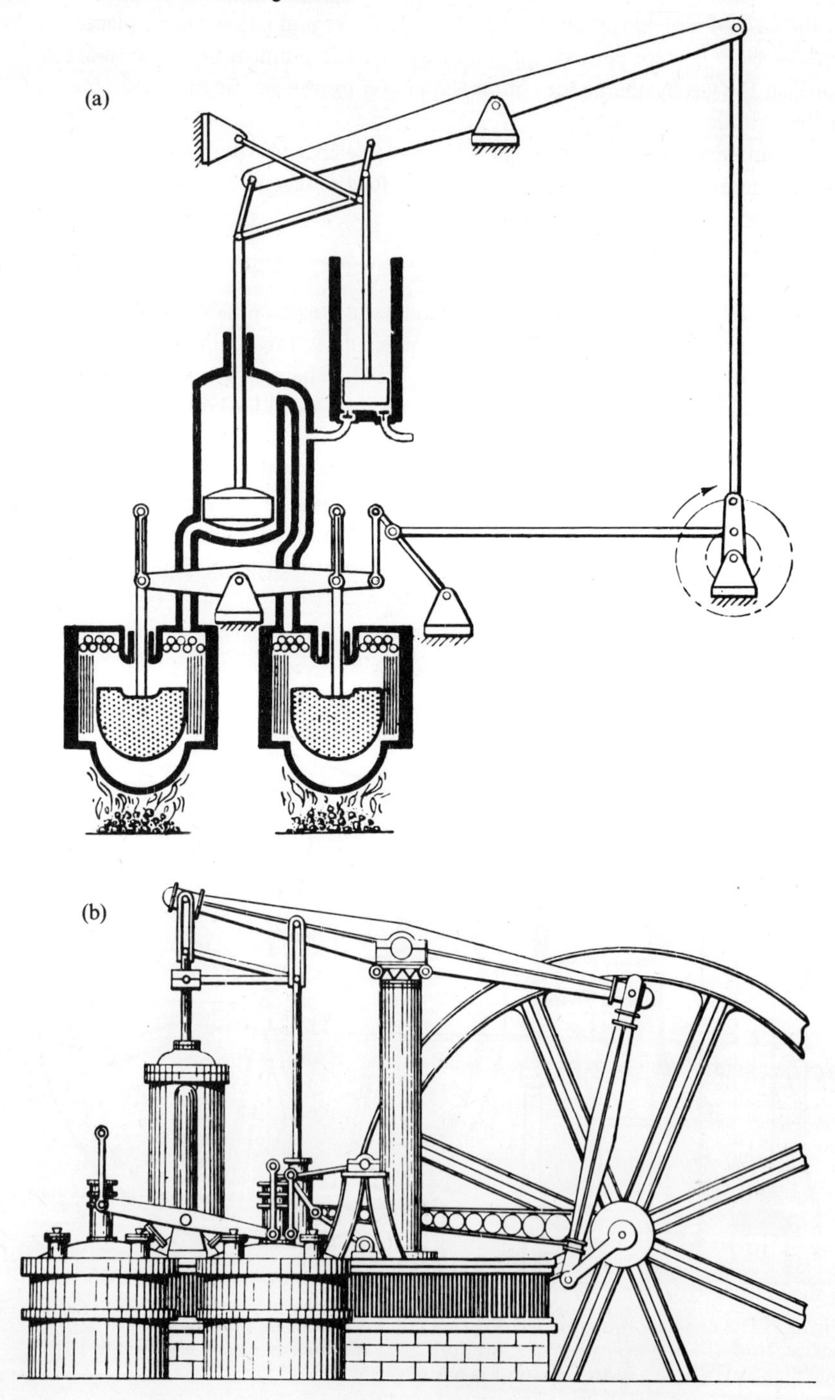
(a)
(b)

have not been developed. The free-piston engine is another interesting configuration. Such machines have recently been brought to an operational stage by Professor Beale of the University of Ohio, and seem promising for future application.

Two-cylinder-per-cycle piston–displacer machines

The first two-cylinder piston–displacer machine was, almost certainly, the double-acting machine, shown in Fig. 6.4, and built in 1827 by Robert and James Stirling. The machine was operated for several years in a Dundee foundry, but was abandoned subsequently, because of repeated failure of the displacer cylinders, caused by overheating.

Single-acting versions of the machine, in various arrangements, are shown in Fig. 6.5. The version having a regenerative displacer may be identified as a Laubereau–Schwartzkopff engine, and, with a separate regenerator, as a Heinrici engine. An arrangement where the cylinder axes were at 90° was made commercially in the last century in fairly large numbers, and was known as the Robinson engine. A machine with interesting possibilities, proposed by H. Rainbow of Bristol, is shown in Fig. 6.5. This machine has two pistons and a single displacer. This arrangement allows considerable flexibility in the drive mechanism, and facilitates the solution of both sealing and cooling problems.

Multiple-piston arrangements

Stirling engines with multiple-piston arrangements can be classified broadly into four groups:

(a) piston–cylinder combinations,
(b) rotary assemblies,
(c) bellows and diaphragm types,
(d) free-piston devices.

Piston–cylinder combinations are the best known of all these groups, and may be further subdivided into single-acting and double-acting arrangements. Fig. 6.6 shows a variety of two-piston machines, three with stationary cylinders and one rotary-cylinder machine. Of these possibilities, only the Rider arrangement of parallel cylinders with the pistons coupled to a crankshaft was produced in any quantity in the last century.

Two forms of double-acting multiple-piston engine are shown in Fig. 6.7. In one arrangement, by Finkelstein, two cylinders (each containing a double-acting piston), are combined to operate as a twin-cycle twin-cylinder engine. One cylinder contains both the compression spaces, and the other both the expansion

---

Fig. 6.4. Early double-acting Stirling engine with piston and displacer in separate cylinders.
(a) Diagrammatic cross-section, showing arrangement of the drive mechanism.
(b) Copy of an old engraving of a beam engine dating back to 1827. (After Finkelstein 1959.)

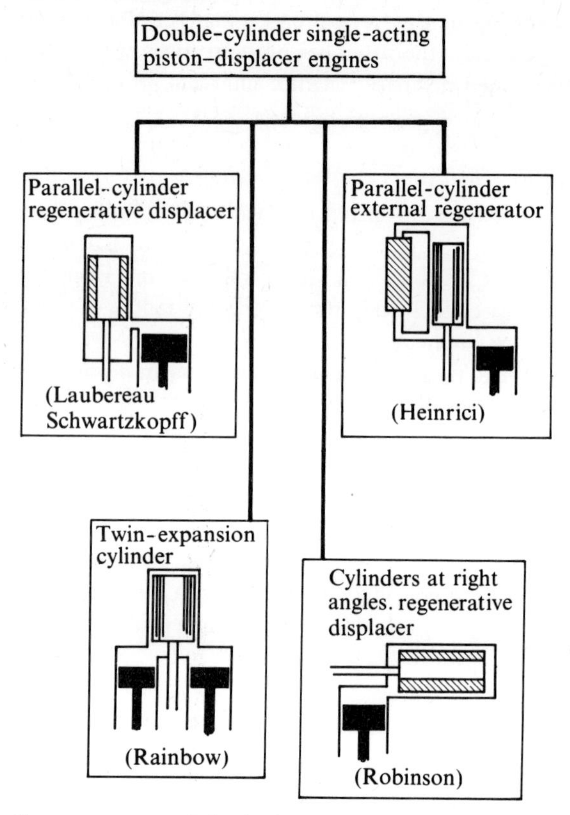

Fig. 6.5. Alternative arrangements of the double-cylinder single-acting piston–displacer engine.

spaces, thereby providing the possibility of complete physical separation of the hot and cold regions of the machine. The other arrangement, by Rinia, may be applied to any number of cylinders, arranged so that the space *above* the piston of one cylinder is connected, through a duct containing the regenerator, to the space *below* the piston of the adjoining cylinder. This arrangement is particularly well suited for a three- to six-cylinder version with the cylinders arranged around a pitchcircle, and with a swashplate or crank drive-mechanism. A multiple-cylinder in-line arrangement is not satisfactory, because the extreme spaces must be connected by a long port. The Rinia engine was invented during the early stages of the Philips programme. It was abandoned subsequently because of lubrication and seal difficulties, but has been resurrected for development as a compact high-specific-output engine, for automotive use. The Rinia arrangement, in common with the Finkelstein arrangement, has the advantage that the number of moving parts is only one per cycle, compared with two in every other case.

The possible configurations embracing rotary assemblies, or bellows and

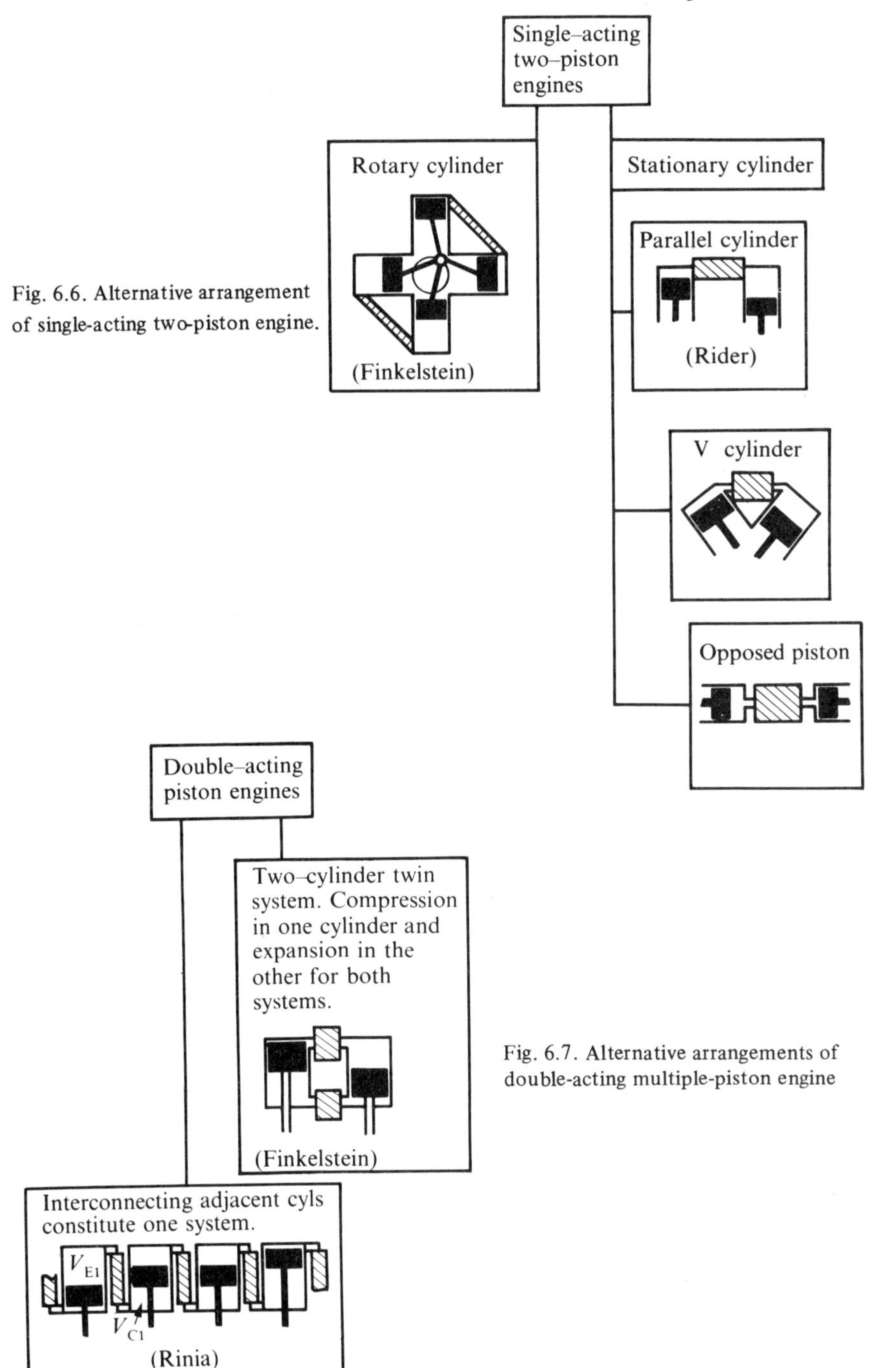

Fig. 6.6. Alternative arrangement of single-acting two-piston engine.

Fig. 6.7. Alternative arrangements of double-acting multiple-piston engine

diaphragm systems, is virtually endless. Most represent attempts to overcome difficulties of imbalance or seal problems, arising from reciprocating elements and the associated linkage, but, so far as is known, none of these have been brought to a commercial stage. A rotary machine, proposed by Zwiauer, is shown in Fig. 6.8. Two Wankel-type rotary engines are coupled on the same shaft, and two regenerators are arranged symmetrically around the axis. One Wankel unit constitutes the compression unit, and the other constitutes the expansion unit. Each unit comprises three distinct spaces, and each space experiences two separate expansion or compression processes per revolution. Thus, a combination of the two engines embraces three separate systems, each undergoing two complete cycles per revolution. It is thought this arrangement could provide a compact high-specific-output machine.

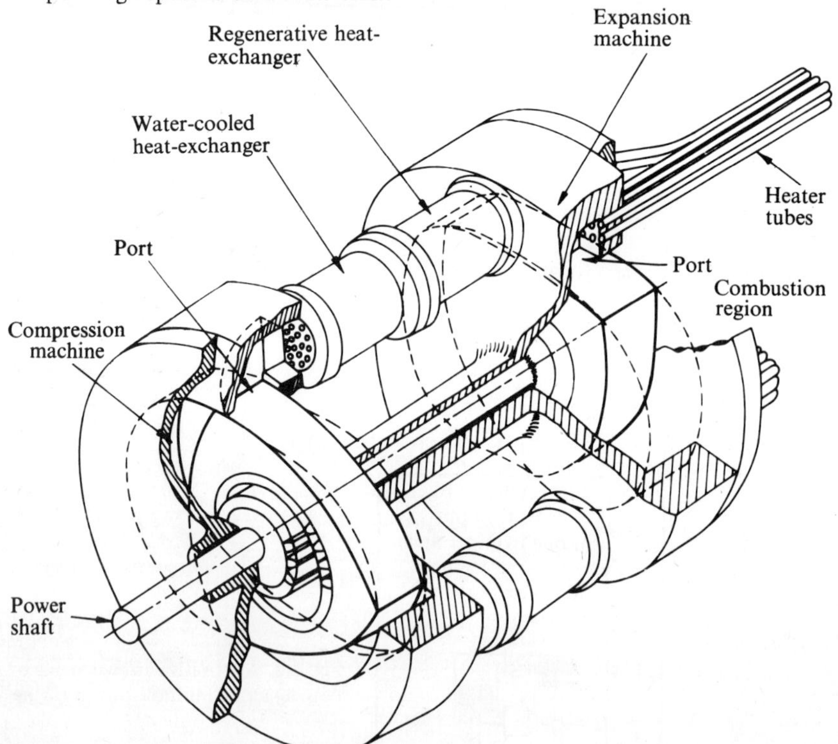

Fig. 6.8. Zwiauer–Wankel configuration of rotary Stirling engine.

Piston–displacer versus multiple-piston engines

It has been shown that many different arrangements of piston–displacer machines and multiple-piston machines are possible, some of which have been developed to a commercial degree. There is no one arrangement that excels above all others in every case, but there are a number of general factors leading to a preferred choice of the single-cylinder piston–displacer machine for most small-size applications:

in larger machines, the choice may lie between multiple single-cylinder piston–displacer machines, on a common crankshaft, or multi-cylinder Rinia or Finkelstein machines.

One important reason for the preference of piston–displacer machines over multiple-piston machines is that, in the latter, it is somewhat easier to deal with the problem of reciprocating seals. On all machines, at least two dynamic-fluid seals are required. In the case of the three machines shown in Fig. 6.1, fluid seals are required on all the pistons, two in the case of the two-piston machine, and one each in the two other piston–displacer machines. An additional fluid-seal is necessary on the displacer rod, as it passes through the piston in one case, and through the underside of the displacer cylinder in the other. The seal around the displacer rod is much smaller than the seal around the piston, with proportionally less leakage and friction. This is, perhaps, the most singular advantage of displacer engines, since the problem of reciprocating seals is particularly difficult, especially when working fluids other than air are used. A further advantage of piston–displacer machines is that the total reciprocating-mass can be less than in multiple-piston machines. This facilitates balancing, and reduces vibration problems. The displacer does no work, and has to withstand merely the gas forces (arising from aerodynamic-flow losses) and its own inertia forces. Therefore, it can be structurally light, and requires correspondingly smaller rods, links, and bearings, so that appreciable savings in weight and mechanical-friction losses can be made.

The power output of a Stirling engine is (to a first approximation), a linear function of the pressure of the working fluid. Thus, one way to immediately increase the specific output is to pressurize the engine. On small engines, it is advantageous to pressurize the crankcase; this not only reduces the duty on the reciprocating fluid-seals, but also reduces the structural strength requirement of the piston and connecting rod assembly, including bearings. This arises from the fact that, with a pressurized crankcase, the pressure difference across the piston is reduced to $(p_{\text{cylinder}} - p_{\text{crankcase}})$, instead of $(p_{\text{cylinder}} - p_{\text{atmosphere}})$ with an unpressurized crankcase. Savings in weight, mechanical-bearing friction and seal friction may be gained thereby. These are offset by the fact that the crankcase is now a pressure vessel with an increased strength requirement, and that at least one dynamic seal is involved if the crankshaft is required to exit from the crankcase.

The problem of a rotating crankshaft-seal is less rigorous than that of a reciprocating piston-seal, and may be eliminated by combining the electric motor, or generator driven by the engine, into the crankcase: however, this may cause appreciable 'windage loss' in a highly-pressurized engine.

Single-cylinder versus two-cylinder piston–displacer machines

It is almost intuitively obvious that, for a pressurized engine, the single-cylinder piston–displacer configuration leads to a crankcase of minimum size and weight. As the engine-power rating increases, the crankcase becomes a dominant fraction of the total engine-weight, and, for large engines, the simple expedient of a pressurized crankcase must be abandoned.

There are two other advantages of the single-cylinder piston–displacer engine over the machine with separate cylinders. In the two-cylinder machine, (Fig. 6.1(b)), the compression space is divided between the displacer cylinder and the piston cylinder, and includes the port connecting the two. This space can never be reduced to zero, so that (in effect), the compression space has a large clearance volume. This clearance volume must be included with the dead space X and, as we have seen earlier, any increase in X results in a decrease in power output.

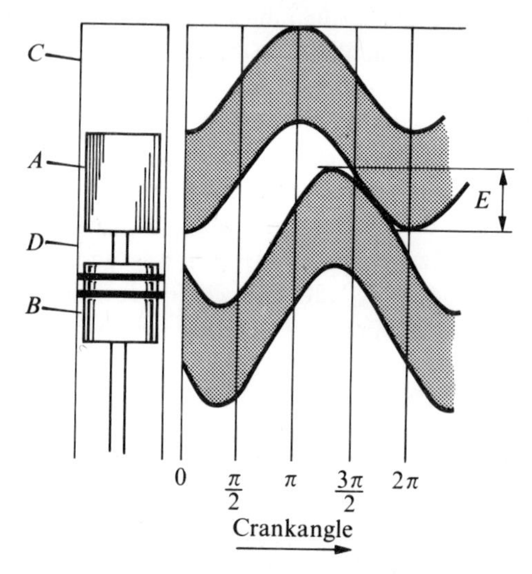

Fig. 6.9. Piston and displacer motion in a single-cylinder engine.
A – displacer, B – piston, C – expansion space, D – compression space, E – section of cylinder traversed by both the displacer and piston.

The second advantage of the single-cylinder piston–displacer engine is that, in every revolution, the displacer and piston both sweep the same part of the cylinder, although at different times. This overlap of strokes is shown clearly in Fig. 6.9, and represents a most efficient utilization of the available engine-cylinder volume.

The advantages of the separate-cylinder piston–displacer machine are:

(1) The increased flexibility for production-engineering design of the crankshaft and connecting rod system,
(2) the separation of the displacer-rod seal to a fixed location in the displacer cylinder, rather than the more limited environment of the piston crown.

In practice, these are very important advantages.

Design variants of Ericsson engines

Regenerative engines of the type where the flow is controlled by valves (called here Ericsson engines) are found, like their Stirling cousins, in a wide variety of types, shapes, and sizes. Sometimes engine arrangements for both types can be very similar; the only distinction between them is the existence (or non-existence) of valves, which allow the passage of fluid through the working space in a cyclic manner, and generally control the flow of the working fluid. In this definition, we exclude gas valves used as part of the control system on Stirling engines to vary the working fluid pressure. This family of machines are not considered here in detail, but it is thought that a brief guide to the principal types might be in order, so as to assist in identification. The degree to which any of the theoretical material, or accumulated engineering experience, here discussed, might be applied to Ericsson engines is unknown.

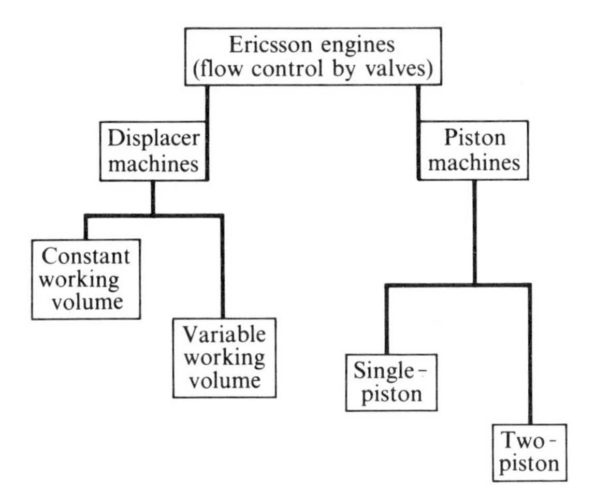

Fig. 6.10. Design variants of Ericsson engines.

Fig. 6.10 is a 'family tree' of Ericsson engines. In most cases, they can be classified either as displacer machines or as piston machines. Each of these principal groups can be further subdivided. Displacer machines may have either a constant working-volume or a variable working-volume. Piston machines may be classified into single-piston machines or two-piston machines.

Fig. 6.11 shows some of the design variants of displacer machines, and identifies some of the better-known arrangements by the names of their inventors. Of the variable working-volume type, only one example is shown. This was first used by John Ericsson, and contains both a piston and a regenerative displacer, coupled together (but moving out of phase) by means of a crank–connecting-rod system. The arrangement may be equipped with gas-operated (or mechanically-operated) valves. This arrangement is potentially attractive for very large nuclear-reactor

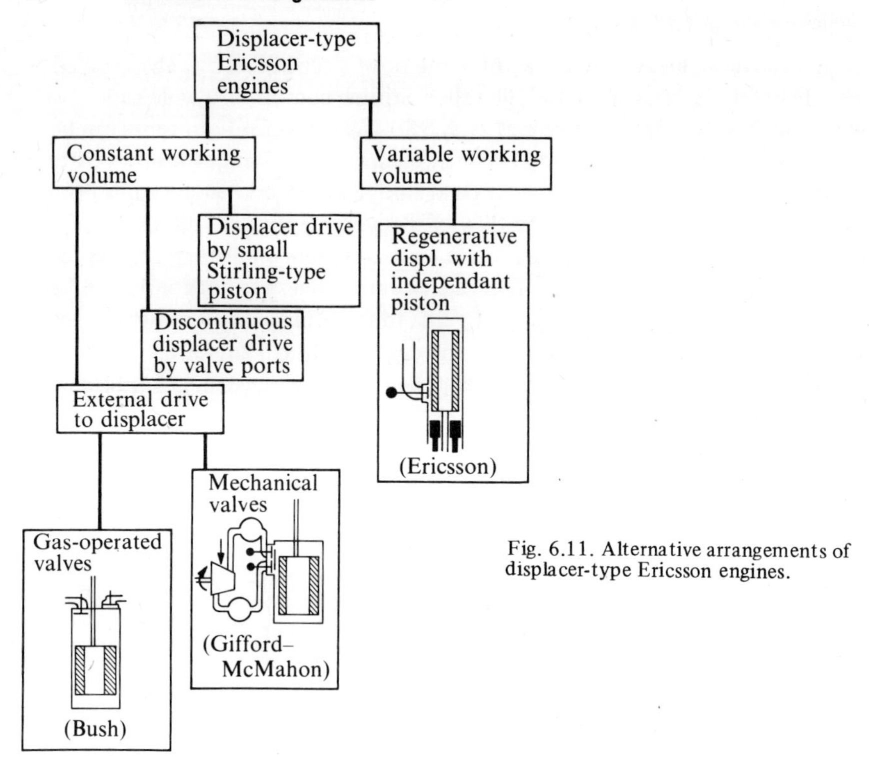

Fig. 6.11. Alternative arrangements of displacer-type Ericsson engines.

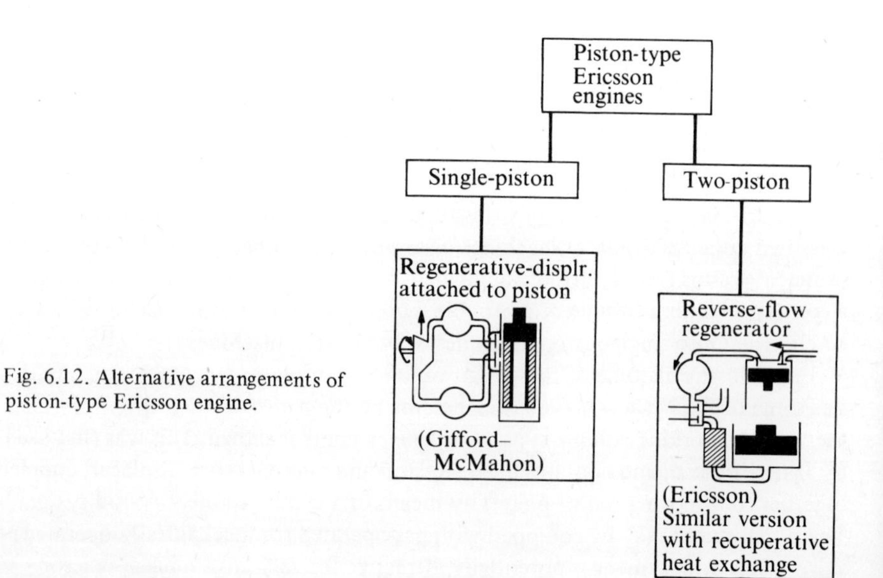

Fig. 6.12. Alternative arrangements of piston-type Ericsson engine.

installations, where the working fluid could be passed as the coolant through the reactor cone.

There is a larger range of possibilities in machines of constant working-volume. Three separate groups are shown on Fig. 6.11, distinguished by the manner in which the displacer is driven. Of the type with an external drive to the displacer, an arrangement with gas-operated valves was patented by Bush for use as a pressure generator, whereas the type with mechanical-operated valves has been successfully developed commercially as a Gifford–McMahon cryogenic cooling engine. Martini (1968) and Buck (1968) have described machines of this type in accounts of their work on the development of implantable artificial hearts. Both refer to their machines as 'Stirling engines'.

In Fig. 6.12, one example each of single-piston engine and two-piston engine arrangements are shown. Many others are possible. The two-piston Ericsson machine may be equipped with a regenerative (or recuperative) heat-exchanger, and is a highly flexibile system.

7 Regenerative heat-exchangers in Stirling engines

Introduction

Designing the regenerator, and other heat-exchangers, of a Stirling engine is a difficult art. So far as the writer is aware there are no well-established procedures closely related to the operation of these components in Stirling engines. Design must be undertaken, therefore, by using fundamental guidelines and a few 'rules of thumb', with the expectation that modifications will be made in subsequent development of the unit. This may appear surprising in view of the great amount of existing literature on the subject of heat-exchangers, including regenerators. However, that such a situation should exist perhaps emphasizes the urgent need for further work, both theoretical and experimental, as an aid to regenerator design.

Here, it may be helpful to consider the problem that confronts the designer of a Stirling-engine regenerator, and to refer to the principal literature. At the end of the chapter, we shall provide some design hints that may be of use, pending the establishment of a more satisfactory approach.

Ideal regenerator

Ideal regeneration was assumed in our previous discussion of both the Stirling cycle and Schmidt cycle of operation. Ideal regeneration is achieved when the fluid entering and leaving the matrix does so at one of two constant temperatures, T_E at the expansion end and T_C at the compression end of the matrix. This is possible only if operations are carried out infinitely slowly, if the heat-transfer coefficient or the area for heat transfer is infinite, or if the heat capacity of the fluid or matrix is zero or infinite, respectively.

In both the Stirling and Schmidt cycles there is no difference in the instantaneous pressure across the matrix, so that the ideal regenerator has no fluid friction. Further, in the case of the Stirling cycle, the void volume of the matrix is zero. In the Schmidt cycle, the void volume is an independently-chosen parameter, and is considered part of the total void volume of the system.

The form of the temperature field in the regenerative matrix is not significant for either the Stirling or Schmidt cycle, but is usually represented as a linear, or transitional, function along the length of the matrix. It is important in the Schmidt cycle, because the effective temperature of the dead space T_D is always taken as the arithmetic mean of the constant temperatures T_E and T_C.

Practical regenerator

The regenerator in a practical engine operates under conditions far removed from those assumed for the ideal case, discussed above. The temperatures of the working fluid at the inlet to the matrix are not constant, but vary with cyclic periodicity, because the processes of compression and expansion are not isothermal. The temperatures at the exit from the matrix are also variable, not only because of the inlet periodicity, but also because limited coefficients of heat transfer and limited surface area, within the matrix, lead to finite rates of heat transfer. The flow conditions at the inlet to (or exit from) the matrix are not constant, but vary continually. The pressure, density, and velocity vary over an appreciable range, and the temperature varies over a more limited range.

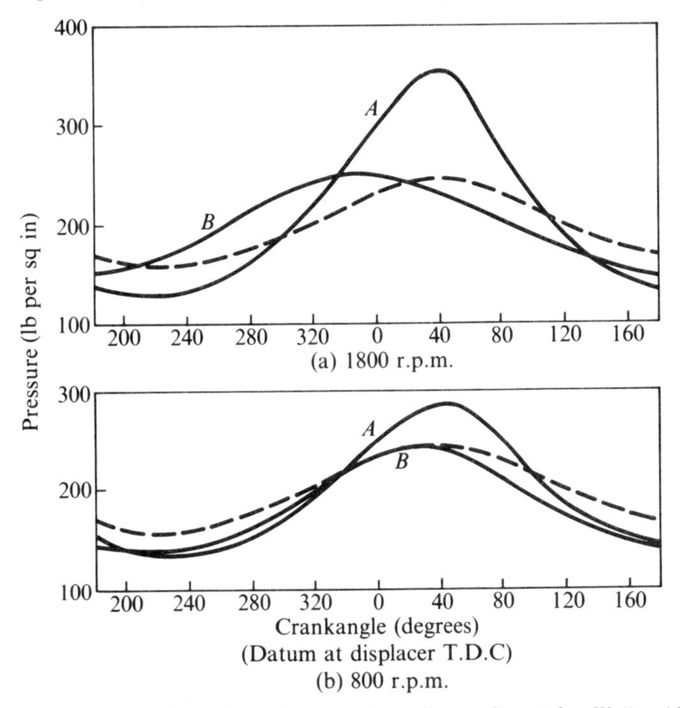

Fig. 7.1. Cyclic pressure variation in a Stirling-cycle cooling engine. (After Walker 1961a.) The figure shows the measured cyclic pressure variation in the expansion and compression spaces of a Stirling-cycle cooling engine operating at speeds of (a) 1800 rev./min and (b) 800 rev/min.†

† On both diagrams, curve *A* is the pressure diagram for the compression space, and curve *B* is the pressure diagram for the expansion space. The difference between them is in the aero-dynamic-pressure loss in the regenerator and heat-exchangers. For comparison, the theoretical pressure diagram, calculated using the Schmidt isothermal theory, is also shown. It is interesting to note that the Schmidt diagram has, approximately, the amplitude of the expansion-space diagram and the phasing of the compression-space diagram. Change in speed causes changes in the amplitude of the compression-space diagram and in the phasing of the expansion-space diagram.

As an example, Fig. 7.1 shows the cyclic pressure-variation in the compression and expansion spaces of a Stirling-cycle cooling engine. The diagrams are reproductions of curves obtained using a Farnborough engine indicator. Two sets of curves are shown for different speeds, and, on both sets, the theoretical Schmidt pressure is given, for purposes of comparison.

Fig. 7.1 shows the *amplitude* of pressure variation in the *expansion space* to be almost exactly that of the Schmidt curve. However, with regard to *phasing* it is the *compression-space* curve that corresponds to the Schmidt curve. Furthermore, it is of interest to note that, as the speed changes from 800 to 1800 rev./min, the principal effects are changes in the *amplitude* of the *compression-space* diagram and in the *phasing* of the *expansion-space* diagram. The difference between expansion-space diagrams and compression-space diagrams results from the fluid-friction loss across the freezer, regenerator, and cooler heat-exchangers. It can be seen that the fluid-friction loss is not negligible, and varies in a rather complex way with speed. A similar relationship was observed with regard to density effects (Walker 1963).

Further calculations, for the same machine, were carried out, assuming it to be operating strictly on the Schmidt cycle. Fig. 7.2 shows the cyclic-mass-flow diagram calculated for two different mean pressures. The diagrams are somewhat complicated. Each diagram contains two curves, superimposed. One curve represents the mass-flow rates into, and out of, the expansion space; the other represents mass-flow rates into, and out of, the compression space. Curves above the zero datum line represent flow into the expansion space, and out of the compression space; curves below the zero datum line represent flow out of the expansion space, and into the compression space. When these are superimposed, the areas where the curves overlap represent the period of net flow through the dead space, including the regenerator. In a cooling engine the 'hot blow' is towards the expansion space, and the 'cold blow' is towards the compression space.

The important point that emerges from a consideration of Fig. 7.2 is that net flow through the dead space (which means, in effect, the regenerator), does not

---

Fig. 7.2. Mass flow in a Stirling-cycle cooling engine. The figure shows the cyclic mass-flow rate for a Stirling-cycle cooling engine, calculated using the Schmidt isothermal theory, for mean pressures of (a) 26 and (b) 11 atm.†

† On each diagram, two curves show the mass flow into, and out of, the expansion space, and the mass flow into, and out of, the compression space. The flow rates were different because the temperatures and swept volumes of the expansion and compression spaces were different, being 70 K, 300 K, 7 in^3 and 11·5 in^3, respectively. In period $A-B$ (hot blow) fluid flowed *into* the expansion space and *out of* the compression space, so that there was net flow through the dead space (regenerator) towards the expansion space. In period $B-C$, fluid flowed into the expansion and into the compression space, so that there was net flow *out* of the dead space. Corresponding flows through the dead space and into the dead space occurred in periods $C-D$ (cold blow) and $D-A$, respectively. Net flow through the regenerator occurs for only a fraction of the total cycle-time.

occur for much more than half the time available. Referring to Fig. 7.2 (upper diagram); from A to B, there is net flow through the regenerator in the direction of the expansion space; from B to C, fluid flows *out* of the regenerator, in both directions, towards the expansion and compression spaces; from C to D, there is net flow through the regenerator in the direction of the compression space, and from D to A, fluid flows *into* the regenerator from both the compression space and the expansion space.

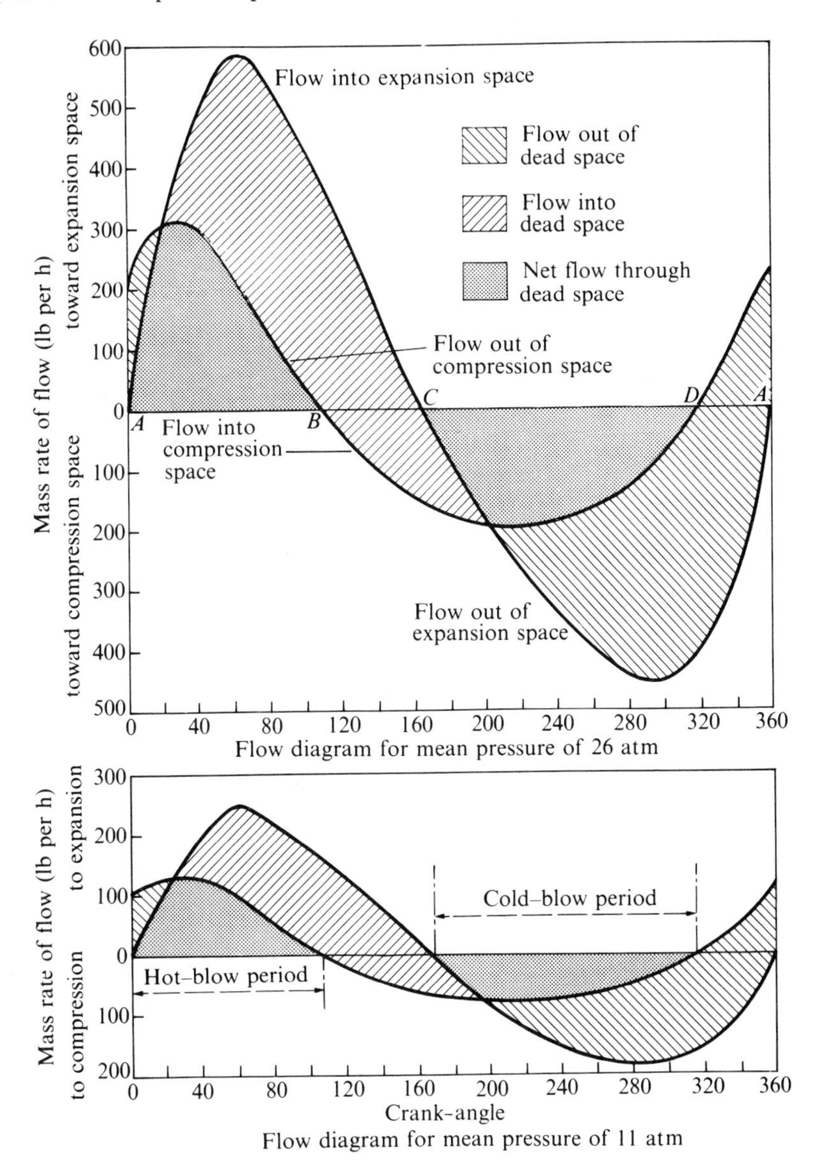

Extending the calculations yet further, it is possible to speculate on the trajectories of particular particles of the working fluid. This has been done in the Fig. 7.3. In this figure, all the volumes have been set adjacent and in sequence, with expansion-space volume variation at the top, and the compression-space volume variation at the bottom. On the volume diagram, the curves at the extremities represent the trajectories of particles which are riding on the piston and the displacer. The diagram also shows cyclic trajectories for a number of other particles at intermediate stations. One of particular interest has been emphasized; *this particle never leaves the regenerator, but oscillates within the matrix throughout the whole cycle.* This shows that none of the fluid ever passes right through

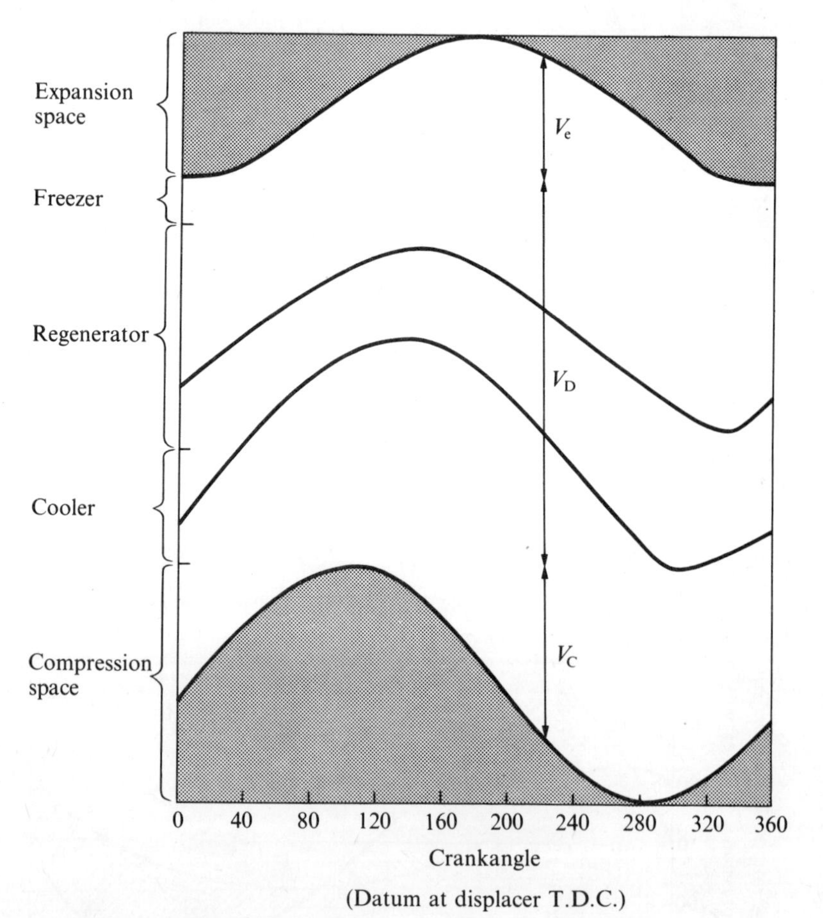

Fig. 7.3. Fluid particle displacement. The figure shows the cyclic trajectory of particular particles of the working fluid in a Stirling cycle cooling engine calculated using the Schmidt isothermal theory. The significant point of interest is that no particle of the working fluid ever passes from the expansion space to the compression space, or indeed ever passes right through the regenerator.

the regenerator. It is an extraordinary result which is of importance when applying classical regenerator theory to Stirling-cycle engines.

The main conclusion to be drawn from Figs. 7.1, 7.2, and 7.3 is that the working fluid in Stirling engines has a complex flow-pattern.

Theory of regenerator operation

The most comprehensive treatment of thermal regenerators is that given by Jakob† comprising a distillation of the classical work of Hausen (1929, 1931), Nusselt (1927), Schumann (1932), and Anzelius (1926). Elsewhere, Iliffe (1948) has reviewed and extended the work of Hausen and others. Coppage and London (1953) have summarized and compared the various results presented in the literature, and Kays and London (1958) have established a rational basis for the design of regenerators, correlated with the design of other forms of compact heat-exchanger. Valuable contributions have been made also by Johnson (1952) and Tipler (1948). None of the work was directed specifically to the application of regenerators in Stirling engines, but was either of a fundamental nature or specific to gas-turbine applications.

Operating conditions

Various modes of regenerator operation may be postulated, but that which is generally of most interest is called the *state of cyclic operation.* This is the state obtained when, after repeated heating and cooling for a fixed time-cycle consisting of one heating and one cooling period, the temperature at any one point in the fluid (or the matrix) is then the same as it was a full cycle earlier.

Fig. 7.4 is a representation of a thermal regenerator in counterflow operation. In the state of cyclic operation the regenerator is assumed to function as follows.

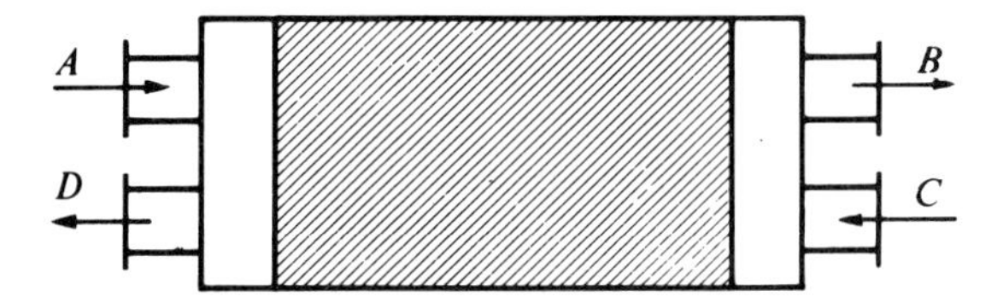

Fig. 7.4. Thermal regenerator in counter flow operation. (a) Hot fluid *A* enters matrix at constant inlet temperature during the hot blow. (b) Fluid *A* leaves matrix at a variable temperature always below the inlet valve, but increasing with time asymptotically to the inlet valve. (c) After the flow of fluid *A* ceases, cold fluid *B* enters the matrix at constant inlet temperature during the cold blow. (d) Fluid *B* leaves the matrix with a variable temperature always above the inlet valve, but decreasing with time asymptotically to the inlet valve.

† Jakob, M. (1957). Regenerators, Chapter 35. *Heat Transfer,* Vol. II, Wiley and Sons, New York.

Hot fluid at a constant *inlet-temperature*, entering from the left-hand end, passes through the matrix, gives up part of its heat, and leaves the right-hand end with a variable temperature, lower than the inlet temperature. The supply of hot fluid is discontinued, and all the fluid is ejected from the matrix through the exit at the right. Cold fluid now enters at a constant inlet-temperature from the right, passes through the matrix, is heated by absorbing heat from the matrix, and leaves at the left-hand end with a variable temperature above the inlet-temperature. The cold fluid supply is discontinued, and all of the fluid is ejected from the cold end, to complete the cycle of operations.

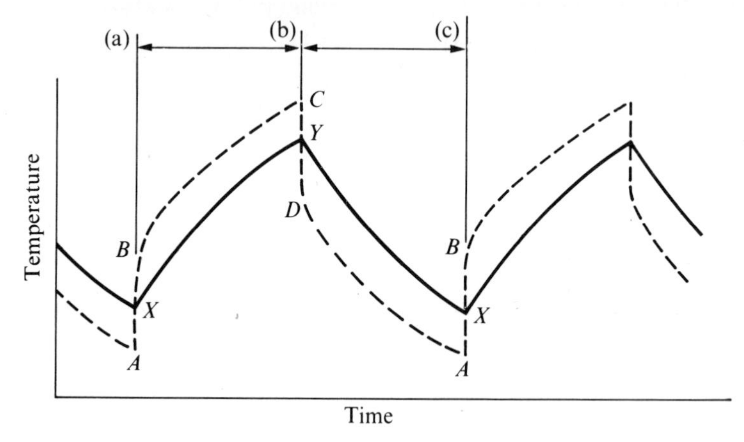

Fig. 7.5. Time–temperature variation of fluid and matrix in a thermal regenerator. The figure shows the possible form of the time–temperature variation at some interim point in a thermal regenerator in the state of cyclic operation. (a) to (b) is the hot blow period . The fluid temperature increases from A to B as the flow is switched from the cold to hot fluid and thereafter increases towards the hot-fluid inlet temperature. The matrix temperature increases from X to Y during the hot-blow period due to heat transferred from the hot fluid to the matrix. At (b) the flow is switched to cold fluid and the period (b) to (c) is the cold-blow period. As the flow is switched the fluid temperature decreases from C to D and thereafter decreases asymptotically towards the constant inlet temperature. During the cold blow the matrix temperature decreases from Y to X as heat is transferred from the matrix to the fluid.

Figure 7.5 shows the possible form of variaton with time of the matrix temperature and fluid temperature, at one particular station in the matrix, with the regenerator in a state of cyclic operation. Fig. 7.6 shows the temperature field in a regenerator, for both fluid and matrix, at the instant of flow-reversal. The upper curves represent the temperature of the fluid and matrix at the end of heating-blow and the start of the cooling-blow. The lower curves represent temperature conditions at the end of cooling-blow and the start of heating-blow. At any particular station along the length of the matrix, the temperatures may fluctuate between the upper and lower curves, in a time-dependent relationship similar to that shown in Fig. 7.5. There are four periods in the cycle. Considering the passage of the hot fluid, the 'blow period' is the time taken for the total quantity of fluid to pass any point in the regenerator; the 'reversal period' is the time which elapses between the entry of one fluid and the entry of the other. Similar blow and reversal periods exist for the passage of the cold fluid. As Iliffe (1948) has pointed out, in practical regenerators the blow period is the same as the reversal period, since the last portion of fluid to enter is driven out by the other fluid through the port by which it came in. In the hypothetical ideal regenerator, the blow period is always less than the reversal period by the time taken for a gas particle to travel from one end of the regenerator to the other. Therefore, if this effect is ignored, we are assuming that the *time for a particle to pass through the regenerator is small compared with the total blow-time.*

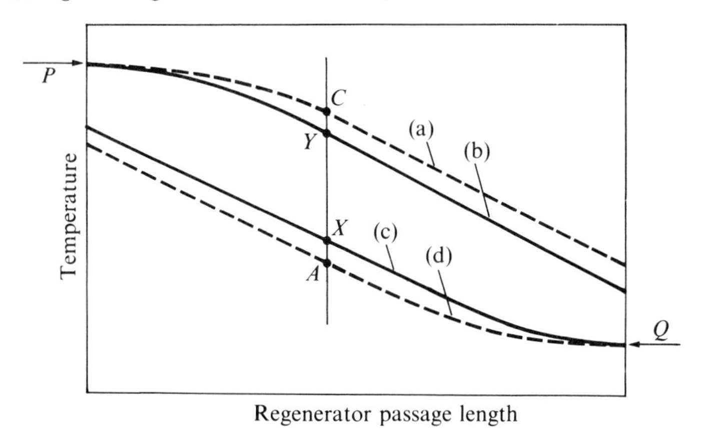

Fig. 7.6. Spatial temperature variation of fluid and matrix in a thermal regenerator. The figure shows the spatial temperature variation in a thermal regenerator, at the instant of flow-switching in the state of cyclic operation, with hot and cold fluids having the constant inlet-temperatures P and Q, respectively.
(a) Fluid temperature at end of the hot blow.
(b) Matrix temperature at end of the hot blow, and start of the cold blow.
(c) Matrix temperature at end of the cold blow.
(d) Fluid temperature at the end of the cold blow.
Points A, C, X, and Y correspond to the conditions represented by A, C, X, and Y in Fig. 7.5.

Other significant simplifying assumptions have been found necessary to render the analysis of operation tractable. Some of these are summarized below.

(a) The thermal conductivity of the matrix must be simple. Nusselt considered four cases:

(i) The thermal conductivity of the matrix is infinitely large. This means there would be no temperature difference in the matrix, and Nusselt's calculation shows that this type would have a poor performance.

(ii) The thermal conductivity of the matrix is infinitely large, parallel to the fluid flow, and finite, normal to the fluid flow. In practice, this may be approached by a very short regenerator, with a matrix composed of thick walls.

(iii) The thermal conductivity of the matrix is zero, parallel to the fluid flow, and infinitely large, normal to the fluid flow.

(iv) The thermal conductivity of the matrix is zero, parallel to the fluid flow, and finite, normal to the fluid flow.

Cases (iii) and (iv) correspond closest to the practical regenerator, but it is unfortunate that the analyses of these two cases are the most complicated. Schultz (1951), Tipler (1947), and Hahnemann (1948) have examined the effect of longitudinal heat conduction in the walls of regenerator passages, and have demonstrated this to have a negligible effect in certain cases. Saunders and Smoleniec (1948) state that 'for matrices built up in layers, such as gauzes, or matrices made of refractory, the conduction effect is almost certainly negligible'.

(b) The specific heats of the fluids and of the matrix material do not change with the temperature.

(c) The fluids flow in opposite directions, and have *inlet-temperatures that are constant both over the flow-section and with time.*

(d) The *heat-transfer coefficients and fluid velocities are constant with time and space,* even though they may be different for the two fluids.

(e) The *rate of mass flow of either fluid is constant during the blow period,* even though it may be different for the two fluids, and the blow periods may be different.

Very little theoretical work appears to have been done on regenerators operating under conditions not fulfilling assumptions (b), (c), and (d), and most results are available for operation with equal blow-times and equal mass flow. However, Johnson (1952) and Saunders and Smoleniec (1948) have investigated this latter effect. Saunders and Smaleniec also considered the effect of variation in the specific heats of the fluid and matrix, for a particular case. They found the assumption of constant values, made in (b), resulted in less than one per cent error in the effectiveness.

Another interesting (but impractical) case, considered by Nusselt (1927), was for a regenerator with an infinitely-small reversal-period, and in which the fluids had been switched infinitely often. The theory for the case is simple, and corresponds to that for a 'recuperator', or normal continuous-counterflow heat-exchanger, in which the two fluids flow continuously, and are separated by metal walls.

Presentation of results

The performance results calculated for regenerators, assumed to be operating under the conditions discussed above, have been presented in a variety of ways. Among the most useful are a series of curves given by Hausen, and reproduced in Fig. 7.7. These have been supplemented by similar curves calculated by

Johnson and Saunders and Smoleniec. The curves show that the effect on regenerator effectiveness of variation in two dimensionless parameters called (after Hausen), the 'reduced length' (Λ) and the 'reduced period' (Π). The reduced length (in the flow direction) is defined by

$$\Lambda = hAL/VC_p,$$

where

h = heat-transfer coefficient between fluid and matrix, per unit surface area,
A = matrix surface area per unit length,
V = fluid-volume flow rate,
C_p = specific heat of the fluid,
L = matrix length.

The reduced period is defined by

$$\Pi = hAZ/MC,$$

where h and A are as defined above,

M = mass of matrix material,
C = specific heat of matrix material,
Z = blow-time.

Frequently Λ and Π are combined by the quotient

$$\Pi/\Lambda = U = VC_p/MC \,.\, Z/L,$$

and called the 'utilization factor', representing the ratio of the sensible-heat capacity of the fluid per blow to the heat-storage capacity of the matrix.

In practice, regenerators may have different reduced periods and reduced lengths for the hot and cold blows, so that there are four factors to be considered. In these cases, Saunders and Smoleniec recommend that average values be used, suggesting (on the evidence of calculations carried out by Johnson), that the error is small. This is probably because, even when the actual blow-times are unequal, the reduced periods are much nearer equality, since a reduction in the actual blow-time Z is usually accompanied by an increase in the rate of fluid flow V.

The usefulness of the concept of two reduced dimensionless parameters and the curves of regenerator effectiveness is limited by the accuracy of the heat-transfer data. This is generally measured experimentally using the 'single-blow' transient technique, first described by Furnas (1932) and, later, by Saunders and Ford (1940), Johnson (1952), Saunders and Smoleniec (1948), Coppage (1952), Rapley (1960), Vasishta (1969), and Wan (1971). In this technique, the matrix is subjected to a flow of hot fluid, entering with a constant inlet-temperature, and the change in the exit-temperature is measured against time. The theory for 'single-blow' operation was first given by Schumann (1932), and may be used to extract, from the measured data, the heat-transfer coefficient relevant to the particular tested matrix. Very careful measurements are required, and there is, in fact, some doubt as to whether this data can be applied to regenerators operating cyclically. A reasonable amount of heat-transfer data is contained in the references given above, but comparison is difficult because several slightly different forms of presentation have been used.

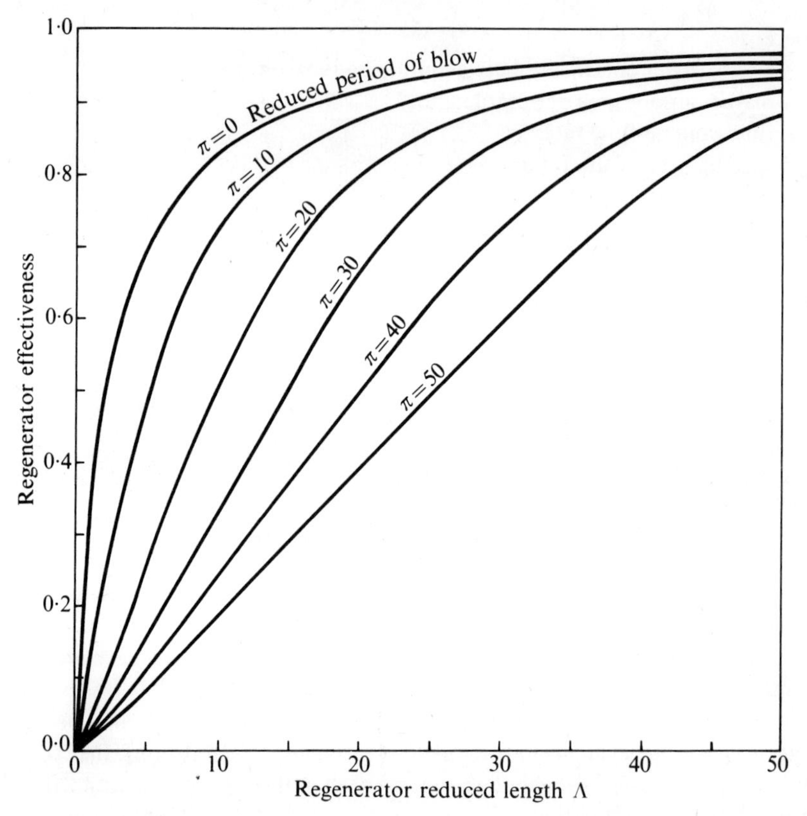

Fig. 7.7. Regenerator effectiveness as a function of the reduced length Λ and reduced period π. (After Hausen.)

Application of theory to regeneration in Stirling engines

Theories of regenerator operation, discussed above, were developed initiallly for air-liquefaction and gas-separation plants and for air preheaters for boilers, These plants are large and, in general, two regenerators per unit are used, one being heated and the other being cooled. The blow-times are very long, ranging from ten minutes to several hours.

Later the theory was adapted and extended, during application of regenerative heat-exchangers to gas turbines. Here the blow-times are much shorter. Coppage and London (1953) refer to 'a reversal-time of a quarter of a second (two complete cycles per second) which is near the maximum permissible frequency without undue "carry-over loss" ' and, again, 'the idealization of no flow-mixing is closely met when the flow-passage length is short, and such shortness of length appears to be good design procedure for the most suitable types of surface'. Most regenerators in gas-turbine engines have a relatively large frontal area and a short flow length, so that, although the blow-time is short, the residence-time of the particle in the matrix is also very short.

The above theory seems applicable, in a reasonably realistic way, to regenerators used in gas-turbine engines and air preheaters, but not applicable to regenerators used in Stirling-cycle engines. The theory is based on assumptions which, clearly, do not apply in the Stirling engine. The most important of these is, perhaps, that the time for a particle to pass through the matrix is small compared to the total blow-time. In a Stirling engine the blow-times are exceedingly short. For example, at the moderate engine speed of 1200 rev./min, or 20 c/s, the blow-time is ten times less than the permissible minimum in a gas turbine. We saw earlier (Fig. 7.3) that the blow-times are so short that no particle ever passes right through the matrix. From Fig. 7.2, we saw the actual net flow-time through the matrix was about half the complete cycle-time, the remaining time being occupied in either filling, or emptying, the dead space. The heat-transfer process that occurs must be very complex, involving a repetitive fluid to matrix, matrix to fluid, fluid to matrix cyclic relationship, rather like the water bucket passed from hand to hand in a fire-fighting operation. Other important assumptions of the theory are that the inlet conditions, temperature, rates of mass flow, and fluid velocity remain constant with time. Clearly, this is not true in *any* Stirling-cycle regenerator. Fig. 7.1 shows that the inlet conditions vary constantly, and Fig. 7.2 shows extreme variation in rate of mass flow. The maximum rate of *net* flow through the matrix is only about half the maximum flow into and out of the expansion space.

Attempts to analyse the regenerators of Stirling engines by any of the recommended procedures require the adoption of 'average' conditions for the flow. Such gross approximation is required to determine these 'average' values that the value of the ultimate result is thought to be highly questionable. No recommendation can be made, at this stage, for the application of any theories of regenerator operation as aids to regenerator design.

Although the situation is unsatisfactory at present, there is reason to hope for improvements. Smith and co-workers at the Massachusetts Institute of Technology have made a promising start (Qvale and Smith, 1968, 1969). They discuss an approximate solution for the thermal performance of a Stirling-cycle regenerator, in which there is provision for non-steady pressure (and mass-flow) conditions, including the possibility of sinusoidal variation, with a phase difference in the peak values. By assuming a second-order polynomial form for the temperature field in the regenerator, a closed solution was obtained for the net enthalpy flux. The theory remains highly idealized, with the assumption that the gas temperature and matrix temperature, at one location, are practically constant with time, and that there are no wall (or fluid-friction) effects. However, at present, the theory does not appear to be sufficiently well developed to be of direct use in regenerator design. Köhler and co-workers at the Philips Laboratories, Eindhoven, have done more research on regenerators in Stirling engines than anybody else, but, unfortunately, little of this work has been published. Although Dr. Köhler presented a series of lectures on regenerators at the Technische Hochschule, Delft, in 1969, these were never published.

In another approach to the problem, the writer is attempting to develop an analysis technique in which the regenerator is subdivided into a number of matrix elements, and the blow-time into a number of periods. The cyclic variation of the flow conditions at inlet is approximated by the assumption of a series of constant conditions for the period, with a step change to new constant conditions for the succeeding period. The exit-conditions from the $(n-1)$th element are averaged to form the constant inlet-conditions for the succeeding nth element. It is possible to increase the number of matrix elements, or blow-time periods, to the point where further subdivision produces no appreciable change. This then represents a stack of Hausen-type regenerators to which the well-developed theory is closely applicable. The method will not be suitable for general design purposes, since it requires access to a high-speed computer. It is intended that a series of consolidated design charts will be generated, to provide for ready selection of an optimum regenerator configuration.

Experimental performance

Little appears to have been published about the effect of imperfect regeneration on the performance of Stirling-cycle machines, or about experimental work on regenerators, tested under conditions approximating to those present in a Stirling engine.

Davies and Singham (1951) carried out some experiments on a small thermal regenerator, composed of brass and copper wire gauzes, subjected to the oscillating flow of a constant volume of air, at atmospheric pressure and at a frequency of five cycles per second. The air was heated on one side of the regenerator, and cooled on the other. Continuous records of the temperature of the air were taken on both sides of the matrix. It was concluded from these experiments that:

(1) for a given gauze matrix, the regenerator efficiency increases with the matrix-weight, but the improvement takes place at a progressively diminishing rate,
(2) for a given matrix-weight, the regenerator efficiency increases with decreases in the diameter of the gauze-wire.

Tests, with equal weights of brass and copper gauze, gave approximately the same values for regenerator efficiency. Thus, although the copper had a thermal conductivity about three times greater than that of the brass, this appears to have had little effect. It was concluded that, with fine wires of these materials, the conductivity lag is extremely small. In these tests, the regenerator efficiency was obtained by analysis of the continuous fluid-temperature records, measured at each end of the regenerator matrix.

Experiments by Walker (1961), with a series of different regenerators on the Philips gas refrigerating machine, have confirmed the second conclusion reached by Davies and Singham, namely, that reduction in wire-diameter increases the effectiveness of the regenerator. The criterion of performance was taken to be the quantity of liquid air produced by the machine, operating at a constant speed and the mean pressure of the working fluid. A reduction in the wire-diameter, with approximately constant matrix-weight and porosity, resulted in an increase in the surface area for heat transfer.

Work by Murray, Martin, Bayley and Rapley (1961) has shed some light on the performance of regenerators under sinusoidal flow conditions. It was found that frequency appeared to have little effect on the heat-transfer process, but the shape of the wave had a significant effect. With pulsating flow, the effectiveness of the tested gauze matrices was appreciably below that obtained under steady flow conditions. With flame-trap matrices, an improvement in the heat-transfer rate, in unsteady flow, was noticed.

Regenerator design—a practical guide

In the absence of adequate theoretical assistance in regenerator design, a few helpful suggestions are offered below. They are not intended to be fundamental rules.

The regenerator designer must attempt to solve the problem of satisfying a number of conflicting requirements. To minimize the temperature-excursion of the matrix, and thus improve the overall effectiveness of the regenerator, the ratio of the heat capacity of the matrix to the gas ($M_m C_{pm}/M_g C_{pg}$) should be a maximum. This can be achieved by a *large, solid matrix.*

On the other hand, the fluid-friction loss must be limited. We saw, in Chapter 3 and Fig. 7.1, that the effect of the pressure-drop across the matrix is to reduce the range of the pressure-excursion in the expansion space, thereby adversely affecting the area of the expansion-space P–V diagram. This reduces the net work-output and thermal efficiency of a prime mover, and the amount of heat lifted and coefficient of performance of a cooling engine. The fluid-friction loss is minimized by a *small, highly porous matrix.*

A third, and most important, consideration is that of dead space. The size of the dead space influences the ratio of maximum to minimum volume of working space, and this directly affects the ratio of maximum to minimum pressure. For maximum specific output, both ratios should be as high as possible, and, for this to be achieved, the dead space should be made as small as possible. This can be achieved by a *small, dense matrix.*

To improve the heat-transfer performance, and establish the minimum temperature-difference between the matrix and the fluid, it is necessary to expose the maximum surface area for heat transfer between the fluid and matrix. Therefore, the matrix should be *finely divided,* with preferential thermal conduction at a maximum normal to the flow, and minimum in the direction of the flow.

Finally, it is important to appreciate that the regenerator acts as an exceedingly effective filter of the working fluid, so that any oil, or grease, particles are retained in the fine flow-passages. In the case of a cooling engine, any impurities in the working fluid that condense in the low-temperature region of the expansion space will accumulate in the regenerator. This build-up is cumulative, and has the effect of increasing the fluid-friction losses, so that the pressure-excursion in the expansion space is decreased, and the performance of the cooling engine progressively diminishes. In the case of the prime mover, any accumulation of oil particles in the regenerator inhibits the flow of working fluid, and increases the pressure loss. The temperature in the expansion space thereby increases, and may be even further increased, because more fuel is supplied in an attempt to restore the lost power. This increase in temperature carbonizes the fuel, thereby further blocking the flow-passage, and the process continues in cumulative fashion, until catastrophic overheating of the engine occurs. From this aspect, the regenerator should offer minimal obstruction to the flow.

Thus, we have the following desirable characteristics for a regenerative matrix:

for maximum heat capacity – a large, solid matrix,
for minimum flow losses – a small, highly porous matrix,
for minimum dead space – a small, dense matrix,
for maximum heat transfer – a large, finely-divided matrix,
for minimum contamination – a matrix with no obstruction.

Clearly, it is impossible to satisfy all these conflicting requirements. With our present understanding of the cycle, it is not possible to quantify the relative significance of the various aspects.

Prime movers

In most engine designs, considerable attention is given to the regenerator, and comparatively little to the problem of the heater and cooler. As a consequence, heat transfer to and from the engine is poor, and the engine fails to operate satisfactorily. This stimulates yet further interest in the regenerator, with the investment of much experimental effort in trying different regenerator-arrangements. Frequently, surprise is expressed when this produces absolutely no effect on engine performance, this becomes confusion when the experiments are extended to the point of diminishing the regenerator to such an extent that it has, in effect, been completely removed from the engine. It is a matter of experience that, in small low-pressure engines, removal of the regenerator nearly always results in improved performance. This is because the gains due to a reduction in the dead space, and, to a lesser extent, a reduction in the conducting path of the regenerator enclosure and in fluid-friction losses, more than offset the loss of thermal capacity and area for heat transfer of the regenerative matrix.

In most small, low-speed machines (up to, say, 5 cm bore, with less than 5–6 atm pressure and operating at below 1000 rev./min), it is adequate (for a start, at least), not to incorporate a formal regenerator into the engine design, but rather to depend on the action of a regenerative annulus around the displacer.

One type of displacer system with a regenerative annular duct, used with success by Professor Beale and also by the writer, is shown in Fig. 7.8. The displacer is made of a thin-walled, low conductivity stainless-steel tube, closed at the hot end by an inverted 'top-hat' section which is machined from a solid bar so as to be a close fit in the tube. After assembly, the seam may be gas-welded, and the joint section trimmed and trued by grinding. Inside the displacer, a series of radiation shields may be provided, as shown, either cut from solid material or fabricated. The lower end of the displacer is closed by another closely fitting plate. Since this end operates in the cooled zone, the end plate can be of light alloy or stainless steel. An epoxy-cement joint has been found adequate for fixing. Good results have been obtained with displacers about three diameters long. The displacer operates in a cylinder, also of low-conductivity stainless steel and having a thin cross-section, except for occasional circumferential stiffening rings, left during manufacture. The top of the cylinder is closed by another inverted 'top-hat' section with an external welded joint. The lower end of the cylinder is, of course, attached by a flange to the cooled compression-space cylinder. The cylinder of the displacer may actually be shorter than the displacer, so that the bottom cooled end of the displacer operates within the compression-space cylinder. This makes it possible to mount a guide ring of P.T.F.E.-based material around the lower end of the displacer, and have it operate on a cooled wall-section.

The annulus formed between the displacer and the cylinder is then the flow-passage, connecting the expansion and compression spaces. It acts as a regenerator, since the top end is always in the heated section, and the bottom end always in the cooled section. It is a simple device, but remarkably effective if the displacer and cylinder wall are reduced to very thin sections, to minimize thermal conduction losses. The gap between the displacer and the cylinder wall is a critical dimension with regard to heat transfer, and should be between 0·015 and 0·030 inches. It is important, also, that a regular annulus be established with a uniform circumferential gap, to equalize heat-transfer and fluid-flow effects. The problem of heat transfer in an annular duct, with an axial temperature-gradient and a

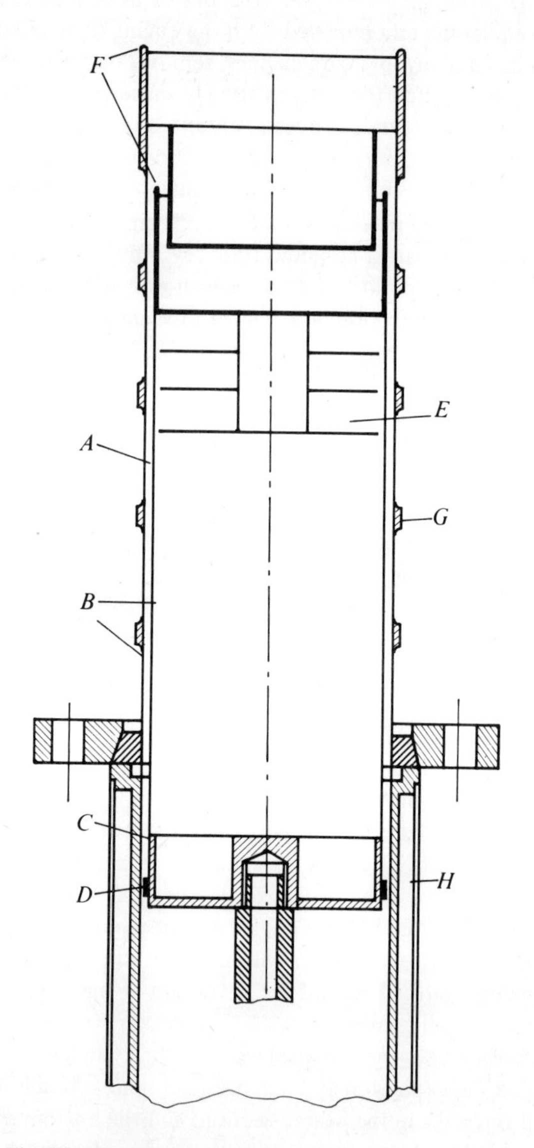

Fig. 7.8. Detail of regenerative annulus.
(a) Regenerative annular gap 0·015–0·030 in.
(b) Thin wall sections to minimize thermal conduction.
(c) Epoxy joint.
(d) Rulon guide-ring.
(e) Radiation shield.
(f) Welded seams.
(g) Stiffening rings.
(h) Cooling jacket.

reciprocating internal member, does not appear to have been studied, and might be a topic of considerable appeal for a university research programme.

The limits of applicability of the regenerative annular duct are not known, but it is likely that the system would becomes less and less effective as the cylinder bore, cylinder pressure, or engine speed were increased. The initial inadequacy would appear, perhaps, in the heater section, and some improvement might be gained by providing an extended surface for heat transfer by using internal finning: but this would be difficult to accomplish without substantially increasing the dead space. Eventually, it would become necessary to resort to increasingly complicated heaters, probably of external-tubular form, and it is at this point that a regenerative matrix becomes worthwhile. By this time, however, one is developing an engine of advanced form that would probably evolve with close similarities to machines of the Philips type.

Cooling engines

The regenerator in a cooling engine appears to be much more important than in an engine. By happy coincidence, the materials problem is less severe than in prime movers.

For their cooling engine, Philips use a regenerator contained in a low-conductivity compressed-paper sleeve, and made up from the random packing of short lengths of copper wire 0·001 inches in diameter, mounted in annular form around the displacer. The author has found woven wire mesh of copper and phosphor bronze to be effective packing for regenerative matrices. These can be had in a wide variety of mesh densities and wire sizes. As the mesh density increases and the wire diameter decreases, the price per unit area increases very steeply to the point where it is doubtful that the material could be used for production machines. An annular regenerator is very expensive, because the centre section, punched from the screen, is 'wasted'. Wire screens can be 'sintered' easily, to form a stable semi-rigid block. One way is to pack the screen in some form that can be loaded with a weight. Then, the wire screen is cleaned by immersion in nitric (or hydrochloric) acid, and the loaded assembly is heated for a short period in a furnace with a reducing atmosphere. On removal, it will be found that the screen has 'sintered' to a solid assembly that can be lightly machined. It is important to arrange the screen so that the wires are normal to the axes of flow, otherwise the axial conduction may be too high. Sintering with light loading does not appear to significantly increase the axial conduction of the screens, because of a considerably reduced porosity, it does improve the pack.

It is not possible to make specific recommendations for design, although the following points merit consideration. The wire used should be fine (0·001 to 0·002 inches in diameter), closely packed, and compressed, to minimize voids. A dead-space ratio of one is a good target, but difficult to achieve, and at least half the dead space should be regenerator void-volume. As a rule, the regenerator arrangement should be of such proportions that the total cross-section of the duct is equal to that of a right normal cylinder, having a diameter equivalent to its length.

Heat-transfer and fluid-friction characteristics of dense-mesh wire screens

The heat-transfer and fluid-friction characteristics of a variety of porous media were given by Coppage and London (1956) and have been supplemented by later data. However, little data on the flow in *dense* wire screens has been published for the size ranges of interest when considering the regenerators of Stirling-cycle cooling engines. Values measured at the University of Calgary by Vasishta (1969), and by Wan (1971), are included here, but no other values are known with which these results may be compared. To validate the experimental apparatus, Vasishta did obtain some results for stainless-steel mesh, in sizes comparable with those studied by Coppage, and found the results to be in close agreement.

The heat-transfer and fluid-friction data for two sizes of screen are given in Figs. 7.9 and 7.10 respectively. Both sizes of screen were woven from phosphor bronze wire, having the following composition:

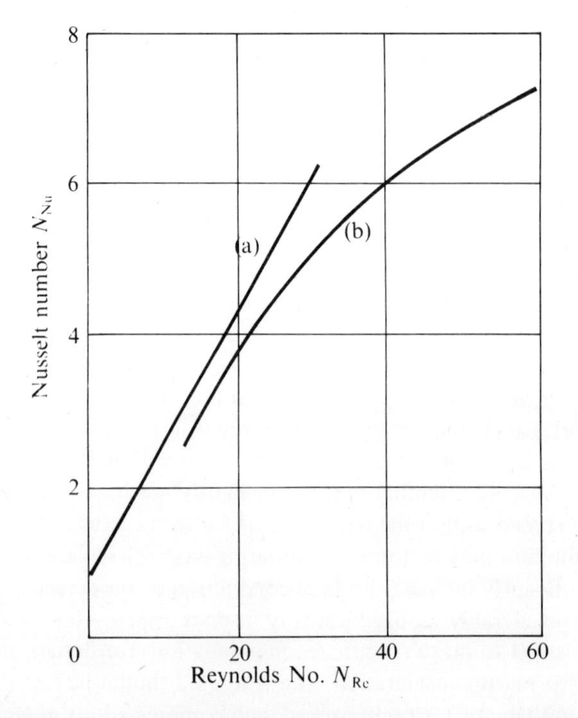

Fig. 7.9. Heat-transfer characteristics of dense-mesh wire screen. (After Walker 1972.)
(a) 400 x 400 strands per inch, 0·001 inch wire diameter.
(b) 200 x 200 strands per inch, 0·002 inch wire diameter.

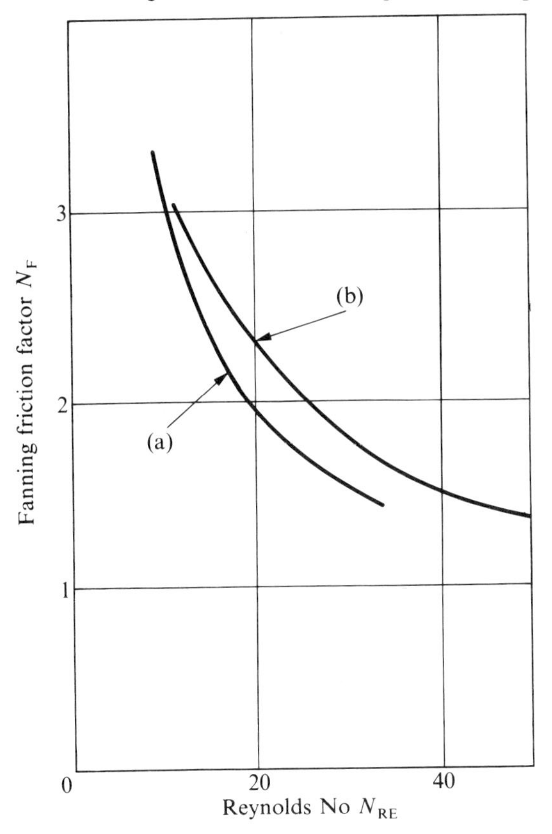

Fig. 7.10. Fluid-friction characteristics of dense-mesh wire screen. (After Walker 1972.)
(a) 400 x 400 strands per inch, 0·001 inch wire diameter.
(b) 200 x 200 strands per inch, 0·002 inch wire diameter.

Tin 3·5–3·8 per cent; Phosphorous 0·3–0·35 per cent; Iron 0·1 per cent;
Lead 0·05 per cent; Zinc 0·3 per cent; Copper rem.
Density 554 lb ft^{-3}; Thermal conductivity 47 B.t.u. $h^{-1} ft^{-1} °F^{-1}$;
Specific heat 0.104 B.t.u. lb^{-1} $°F^{-1}$.

The two screen-sizes investigated were

(a) 200 x 200 strands per inch, 0·0021 in wire diameter,
(b) 400 x 400 strands per inch, 0·001 in wire diameter.

The heat-transfer characteristics are presented as the Nusselt number N_{Nu}, as a function of the Reynolds number N_{Re}, defined as follows:

$$N_{Nu} = (4r_h/k)(h/f), N_{Re} = \rho_f Vd/\mu_f = (4r_h/\mu_f p)(W_f/A_f),$$

where r_h = calculated hydraulic radius of the screen, h = heat-transfer coefficient, k_f = thermal conductivity of the fluid, ρ_f = density of fluid, V = volume flow rate of fluid in matrix, W_f = mass flow rate of fluid in matrix, A_f = frontal area, p = calculated porosity, μ_f = dynamic viscosity of fluid, where p = (volume of matrix – volume of metal)/volume of matrix and r_h = total volume of connected void spaces/total surface area = volume of matrix x porosity/total surface area.

The fluid-friction characteristics are presented as the Fanning friction factor N_f, as a function of the Reynolds number N_{Re}, defined as follows:

$$N_F = 2\rho_f \,.\, \Delta P \,.\, r_h \,.\, p^2 / nLG_A{}^2,$$

where ΔP = pressure drop, n = number of layers of screen, L = length of matrix, G_A = mass flow per unit area, and the remainder are as defined above.

8 The Philips programme

Details of the Philips Company work on Stirling engines have been given in various publications (see Bibliography). The following is a brief summary drawn from this material.

Early history

Work on Stirling engines started in about 1938 at the Philips Research Laboratories, Eindhoven, Holland. It has continued unabated since then.

The initial objective was the development of a small thermally-activated electric-power generator for radio and other small electrical equipment, for use in undeveloped areas, where kerosene and petrol were available. A one horsepower engine, far in advance of anything then known, was developed. The engine was the single-cylinder piston–displacer engine, shown in Fig. 8.1. A similar engine was incorporated in the 200-watt air-cooled engine–generator set, shown in Fig. 8.2. A pre-production run of about 400 of these machines was produced, and disposed of, several years later, to universities and technical colleges, for use as teaching aids. These units are often the only Philips Stirling engines that many have seen (Rinia and du Pre 1946, de Bray *et al.* 1948, and Van Weenan 1948).

Improvements in radio valves, batteries, and, above all, the invention of the transistor later removed the need for small electric-power generators. However, by this time, sufficient work had been done to demonstrate the possibilities of the engine in other applications. Research was undertaken, therefore, on larger engines and (separately) on Stirling-cycle cooling engines.

Prime movers

The engine configuration of Fig. 8.1 was not suitable for large machines, because the need for a pressurized crankcase resulted in a large, heavy machine. The invention of the Rinia engine appeared to overcome this difficulty for, in this arrangement, it is not necessary to pressurize the crankcase.

A diagram of the Rinia arrangement is shown in Fig. 8.3. Multiple cylinders are interconnected, so that the upper expansion space of one cylinder is connected to the lower compression space of the adjacent cylinder by means of a port containing heater, regenerator, and cooler. In the Rinia engine, the number of reciprocating parts was reduced to one per cycle and, in combination with a swashplate drive mechanism, a very compact four-cylinder engine was achieved.

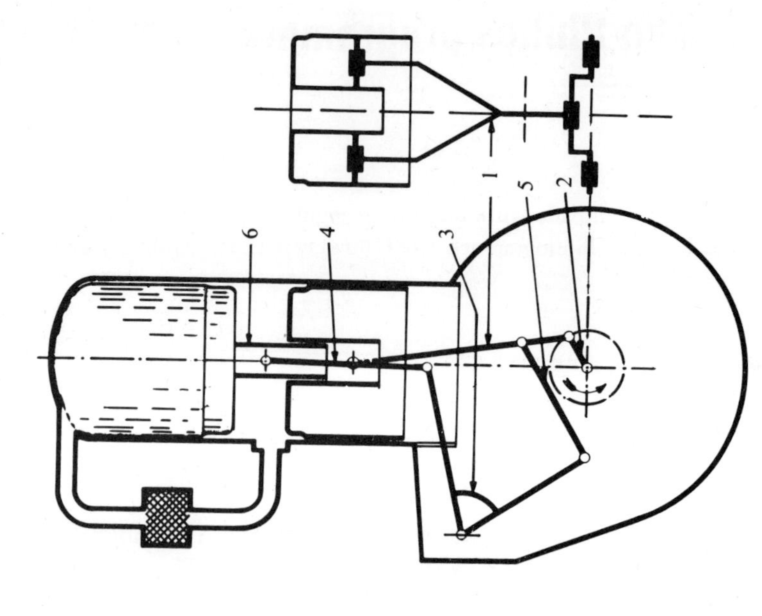

Fig. 8.1. Early one-horsepower air engine (*circa* 1946). 1. Forked piston connecting rod; 2. crankshaft; 3. rocker linked by arms 4 and 5 to the displacer piston rod 6 and to a certain point on 1. (Courtesy of Philips Technical Review.)

Subsequent difficulties (principally in lubrication, sealing, and aerodynamic design), caused the development to be put aside in favour of a return to single-cylinder piston–displacer engines.

Fig. 8.2. 200-watt electric-power generator with air-cooled engine. (Courtesy Philips Technical Review.)

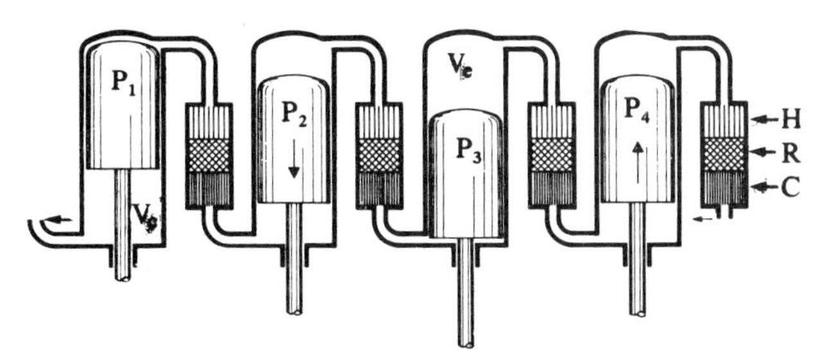

Fig. 8.3. Mechanical arrangement of Rinia double-acting engine. (Courtesy Philips Technical Review.)

This return to piston–displacer machines followed the invention of a new drive mechanism, called 'the rhombic drive', by Dr. R. Meijer (1959). Thereafter, all research and development was concentrated on machines of this genre until, quite recently, there has been a resumption of work on Rinia machines. The invention of rhombic drive for single-cylinder piston–displacer engines and the Rinia configuration for multiple-cylinder machines are thought to be the two important fundamental contributions made to the research by the Philips Company. The rhombic drive allowed a return to piston–displacer engines, without the need to pressurize the crankcase. Furthermore, it provided the possibility for the engine to be completely balanced, a feature not possible in single-cylinder machines with a crank–connecting-rod mechanism. Another advantage of rhombic drive was that friction and wear, caused by side-thrust of the piston and displacer crosshead, was eliminated. The principal disadvantage of the rhombic drive was its relative complexity.

Fig. 8.4 is a diagram showing the elements of a Meijer rhombic drive. It was comprised of twin crank–connecting-rod mechanisms, identical in design, and offset equidistant from the central axis of the engine. The cranks rotated in opposite directions, and were connected by identical timing gears. The connecting rods were coupled by an upper and lower yoke, with the piston attached to the upper yoke, and the displacer attached to the lower yoke. When operating, the piston and displacer performed cyclic motion, near sinusoidal in form, but appropriately different in phase, so that the necessary volume variations in the compression and expansion spaces were achieved.

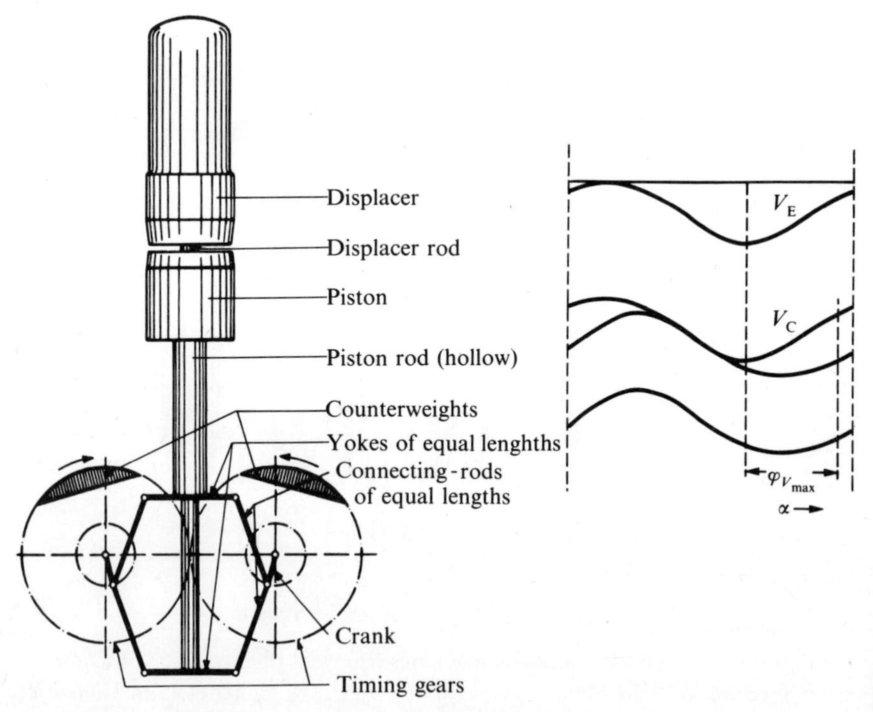

Fig. 8.4. Mechanical arrangement of Meijer single-cylinder piston–displacer engine with rhombic drive. (Courtesy Philips Technical Review.)

Much effort has been devoted to development and refinement of Meijer engines. A range of engines from single-cylinder five-horsepower machines up to four-cylinder 360-horsepower engines have been evaluated experimentally.

Theoretical studies have been made of very large engines, suitable for marine or nuclear reactor applications (Meijer 1969, 1970). Fig. 8.5 is a cross-section diagram of the Meijer engine, at an advanced stage of development. The engine shown is water-cooled, and arranged for heating (by combustion) of fossil fuel.

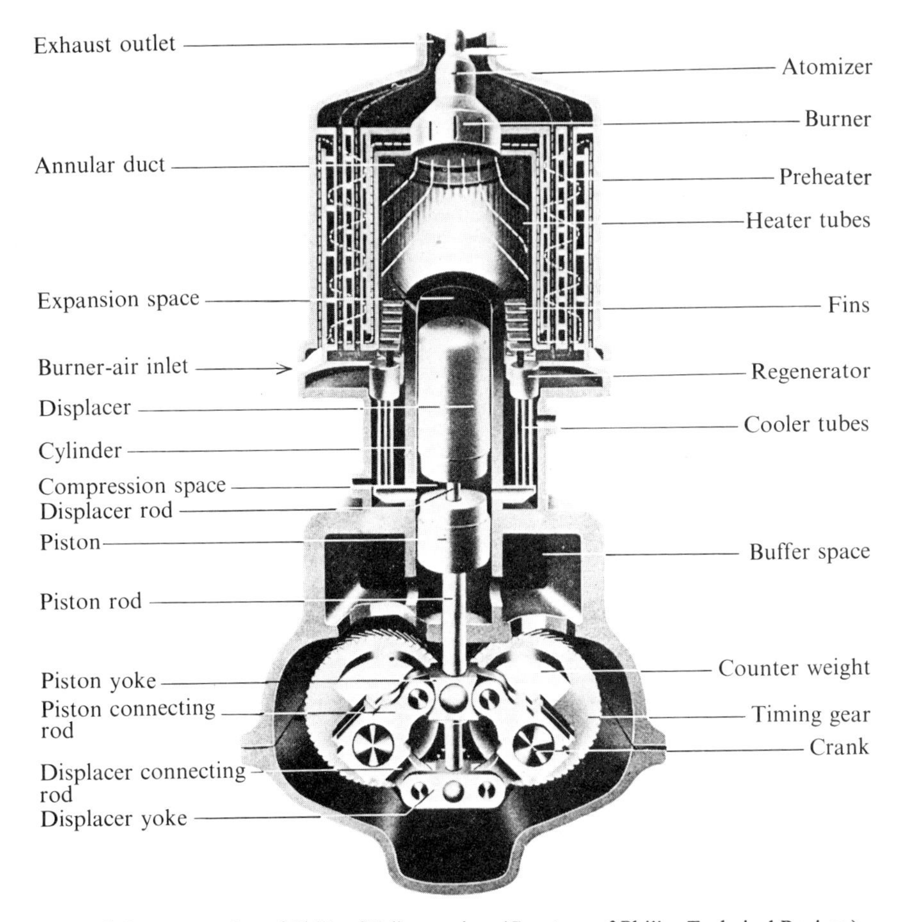

Fig. 8.5. Cross-section of Philips Stirling engine. (Courtesy of Philips Technical Review.)

Advanced engines of this type use hydrogen or helium as the working fluid rather than air, and work at very high pressures. This type of working fluid is used, because the transport properties (i.e. specific heat, thermal conductivity, and viscosity) of hydrogen and helium are most favourable for engines of high specific output.

Fig. 8.6 presents the results of comparative studies made by Philips of the optimum thermal efficiency, as a function of engine specific power, with air, helium, and hydrogen as the working fluid. At low speed and low specific power, the difference is not very great, but becomes increasingly marked at higher values of speed or specific power. It is important to note that the studies refer to large engines of 225 b.h.p. per cylinder. The advantages of hydrogen and helium, although significant, might not prevail to the same degree for smaller engines.

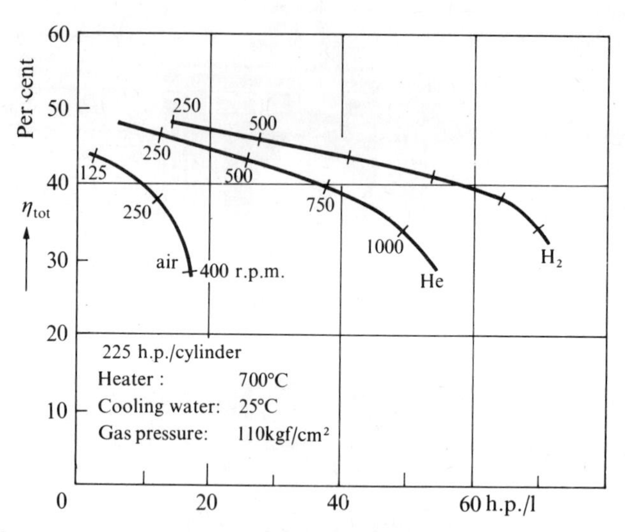

Fig. 8.6. Comparative performance of air, helium, and hydrogen as the working fluid in Philips Stirling engines. The maximum efficiency η of a single-cylinder Stirling engine with rhombic drive and an output of 225 hp, as a function of the specific power (shaft horsepower per litre of cylinder volume) in the case of three different working gases. A higher specific power requires a higher speed and leads to a fall in η. (Courtesy Philips Technical Review.)

With hydrogen or helium as the working fluid, sealing problems become critical, and occupy a substantial amount of research-time. In a Meijer engine, there were, in fact, two distinct sealing problems; firstly, to seal the piston and displacer rods, to prevent leakage of working fluid to the crankcase; secondly, to seal the piston against leakage from the working space to the buffer space, and, to a lesser extent, to seal the displacer. Solutions to these problems were found with the development of the hydraulically-supported roll-sock seal (Fig. 8.7), for the piston rods, and the development of unlubricated Teflon-based sealing rings for the piston and displacer.

Many refinements have been made to heat-exchangers, particularly the heater and the exhaust-gas heat-exchanger, where high effectiveness is required to reduce the 'stack loss' to a small percentage of the supplied heat. Another important development was in the method of engine regulation. Change in the fuel supply did eventually change the temperature regime of the engine, but the change is slow, because of high thermal inertia. A near-instantaneous response was obtained with the development of a control system to regulate the pressure of the working fluid in the cylinder, including a buffer space–cylinder-bypass valve for engine braking.

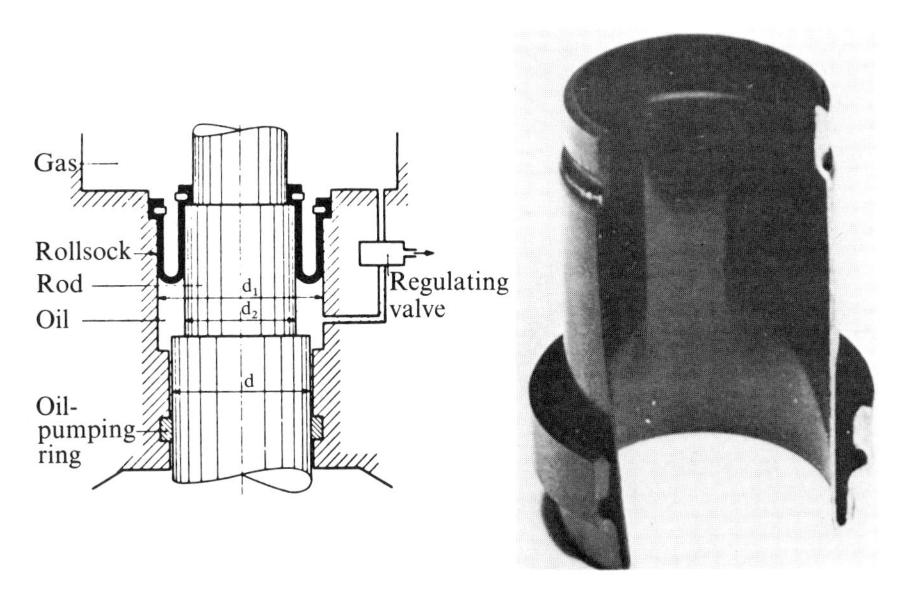

Fig. 8.7. The hydraulically-supported roll-sock seal. (Courtesy Philips Technical Review.)

Increasing public concern with air pollution has focused the attention of Philips engineers on to the possibility of applying the Stirling engine to automotive use. Air-pollution characteristics of the engine compare very favourably with gas-turbine and diesel engines, as shown by the comparative figures (given in Table 8.1) for the quantities (in milligrams per second per horsepower of CO, C_xH_y, and NO_x) exhausted to the atmosphere, at full-load conditions. The virtual absence of carbon monoxide and unburned hydrocarbons in a Stirling engine exhaust is due to the fact that combustion takes place continuously in a hot-walled chamber, and air may be supplied in any excess quantity required, thus eliminating unburned residual gas. It is not obvious

TABLE 8.1. Comparative rates of emission products per horsepower. After Meijer 1969.

Compound $(\text{mg s}^{-1} (\text{h.p.})^{-1})$	Stirling engine	Gas turbine	Diesel engine
CO	0·1–0·3	2·0–3·6	0·2–5·0
C_xH_y	0·003–0·006	0·036	0·6–12·0
NO_x	0·7–0·02	0·7–2·0	0·4–2·0

why the nitrous and nitric-oxide production is reduced, because the flame-temperatures of the burner are relatively high, but even this low value can be reduced by as much as 60 per cent by the relatively simply expedient of recirculating about one-third of the exhaust gas back through the combustion zone.

Other favourable characteristics of the engine for traction applications are the following (Neelen, Ortegren, Kuhlman and Zacharias 1971).

(1) The engine is without vibration, and very quiet compared with diesel engines by some 20 to 40 dB – this is because there are no valves or periodic explosions, and perfect balance can be achieved.
(2) The efficiency and specific output are comparable to diesel engines, with very good efficiency at part-load conditions.
(3) Engine braking is possible with a maximum negative torque up to 80 per cent of full-load torque.
(4) The engine has a wide speed-range, and favourable torque characteristics, which allows for the use of a simplified transmission.
(5) Oil consumption is virtually nil, and very infrequent oil changes are necessary.
(6) Rapid acceleration and response are possible.
(7) The engine has a multi-fuel capability for any liquid or gaseous fuel.
(8) It is reliable, and has a long service life.
(9) It is not susceptible to contamination or damage from dust or salt atmospheres.

Fig. 8.8 is a four-cylinder 200-horsepower engine under development for automotive use, and Table 8.2 is a summary of the principal technical features of the engine. A prototype unit has been installed in a city bus, for evaluation and demonstration purposes, and one Philips licensee (United Stirling of Malmo, Sweden), may start production of this type of engine in 1974, with quantity production in 1976. The price of the engine has not been given, but it is generally conceded that it will be greater than that for a diesel engine of comparable power. In city buses the cost of the engine is about 10 per cent of the total cost, and it may be that the advantages enumerated above will be so great as to allow an increase in the cost of the *bus* by 10 per cent, thereby providing for the cost of the engine to be twice that of the diesel engine, a margin agreed by all to be generous.

A succeeding generation of engines are at an advanced stage of development. These are Rinia engines, highly pressurized and of compact design, so that specific output comparable to petrol engines and thermal efficiencies comparable to diesel engines may be gained. A diagram of an engine of this type is shown in Fig. 8.9. This machine is shown equipped for heat supply by a system of indirect heating, using the 'wick-thermosyphon' or 'heat-pipe' principle. This has been adopted by Philips to overcome the problem of attaining the very high heat-flux required at the heater tubes in compact highly-pressurized engines.

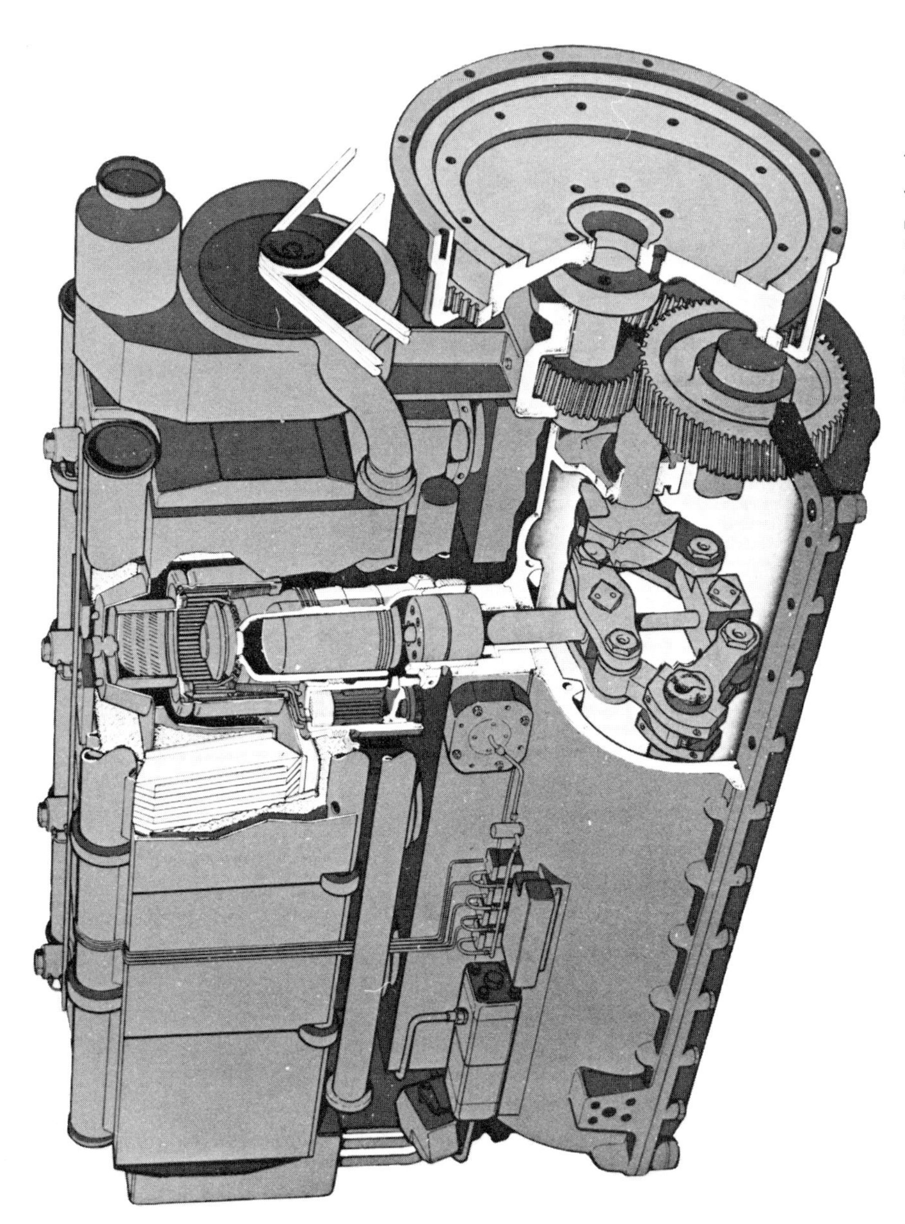

Fig. 8.8. The Philips 4/235 four-cylinder in-line Stirling engine. (Courtesy Philips Technical Review.)

TABLE 8.2. Summary of technical features of the Philips 4/235 Stirling engine.

Type	4–235
Number of cylinders	4
Combustion system	External
Fuel	Diesel fuel
Bore and stroke	77·5 mm dia. x 49·8 mm
Total piston displacement	940 cm^3
Working medium	Helium
Maximum output	200 b.h.p. at 3000 rev./min} final rating
Mean pressure of medium	220 atm
Maximum output	100 b.h.p. at 3000 rev./min} preliminary rating
Mean pressure of medium	110 atm
Normal temperature	700 °C
Normal radiator temperature	60 °C
Maximum torque	35 kgf.m at 1000 rev./min
Efficiency	approx. 30 per cent at 100 h.p.
Lubricating system	Dry sump with scavenge pump
Oil filter	Bypass, replaceable element
Cooling system	Water-cooled with centrifugal pump
Dry weight	760 kg
Dimensions	1250 mm long x 1100 mm high

The heat-pipe is a device for transporting heat at near isothermal conditions, and at a thermal flux density several thousand times better than, say, a pure solid copper bar. It consists of a hermetically-sealed chamber, with a porous lining to the inside walls, called the 'wick'. The chamber contains a fluid, which vapourizes in the hot region and condenses in the cold region, so that there is a flow of *vapour* from the hot to cold regions, and a flow of *liquid* back through the wick. The heat flow from the hot to cold region is very high, because of the latent heats involved in evaporation and condensation. The heat-pipe allows heat to be *absorbed* from the combustion chamber, or other thermal source, at a *low flux density*, but with virtually unlimited extent, since any area for heat transfer may be involved, so that very large amounts may be received *in toto*. At the same time, the heat may be *supplied* to the engine by the heat-pipe at very *high flux densities*, and at virtually the same temperature as that of the heat received by the heat-pipe. At the temperatures of interest in Stirling engines (700 to 800 °C), sodium is a suitable fluid for the heat-pipe.

Use of the heat-pipe has removed the danger of localized hot spots in the highly-pressurized engine heater tubes, thereby permitting an increase of 50 to 75 °C in the maximum cycle-temperature, with consequent improvement in power and efficiency. At the same time, the heat-transfer surface of the heat-pipe which is subject to the combustion gases can be very large, so that the burner-efficiency can be improved. It is possible, also, to have a lower combustion temperature, which allows for a drastic reduction of NO_x emission products in the exhaust.

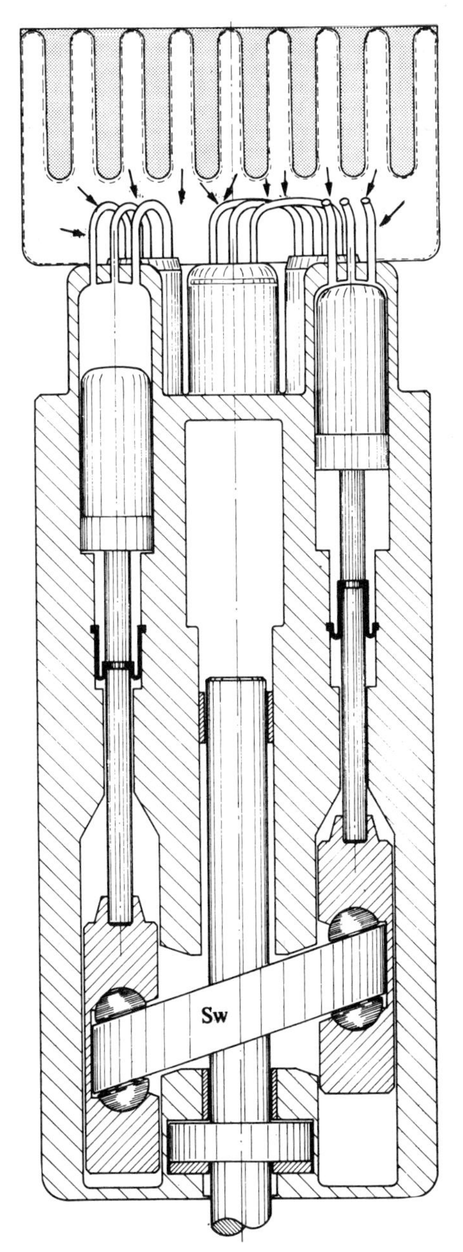

Fig. 8.9. Advanced Rinia double-acting engine with swash-plate drive and indirect heating. (Courtesy Philips Technical Review.)

Stirling engines receive heat indirectly, by transfer through the heater walls, thus permitting virtually any thermal energy source to be used. One possibility is that a thermal storage battery could be charged overnight by electrical heat energy, and discharged by daytime use of a Stirling engine, converting the stored heat to work for driving a vehicle. Philips have investigated this possibility, using lithium fluoride as the thermal storage medium. They have found the system suitable for adaptation for commuter cars, city buses, taxis, and delivery vans. In many respects, it is superior to the alternative of electric vehicles with storage batteries (Meijer 1970).

Another concept being studied concerns the use of hydrogen as the fuel in a combustion-heated Stirling engine. The advantage of hydrogen fuel is that the combustion-product is water, so there are no problems of exhaust emission. The main difficulty with regard to the use of hydrogen as the fuel in vehicle engines has been the problem of storage. A recent discovery at Philips is that some hexagonical intermetallic compounds, containing a rare earth metal and nickel or cobalt, absorb and desorb large amounts of hydrogen at pressures of a few atmospheres. At 2·5 atmospheres pressure and room temperature, the density of hydrogen in $La\ Ni_5$ is almost twice that of liquid hydrogen. If this discovery can be brought to commercial application it may have a significant effect on the problem of air pollution by vehicle engines, generally. Hydrogen may be used as a 'clean' fuel in conventional internal-combustion engines, just as readily as in Stirling engines (Meijer 1970).

Cooling engines

In the course of the early work on prime movers, Rinia and du Pre observed that the engines worked well as refrigerating machines, when driven by an electric motor. In 1945, a one-horsepower machine, operating as a cooling engine, attained a temperature of 83K. Following this, further research on Stirling-cycle cooling engines was assigned to a separate group, led by Dr. J. W. L. Köhler (1955*a* and *b*, 1960). The succeeding twenty five years has seen the rise of the Cryogenerator Division of Philips to a major position in the cryogenic engineering industry. Cryogenic engineering is the name given to engineering operations at temperatures below about 100K, where the so-called permanent gases liquefy (liquid methane, oxygen, nitrogen, argon, hydrogen, helium, etc.).

Köhler has written that he designed his first models along the lines of hot-air engines and, in 1950, obtained temperatures sufficiently low to liquefy air. By 1954, the liquefier shown in Fig. 8.10 was in production, with a liquefaction-capability of seven litres of liquid air per hour. The machine was a single-cylinder piston–displacer engine, driven by an electric motor. The cylinder-head was cooled by expansion of the working fluid (hydrogen or helium) in the expansion space. Atmospheric air was progressively cooled to the point of liquefaction on the cold head, in passage through a large heat-exchanger, in good thermal

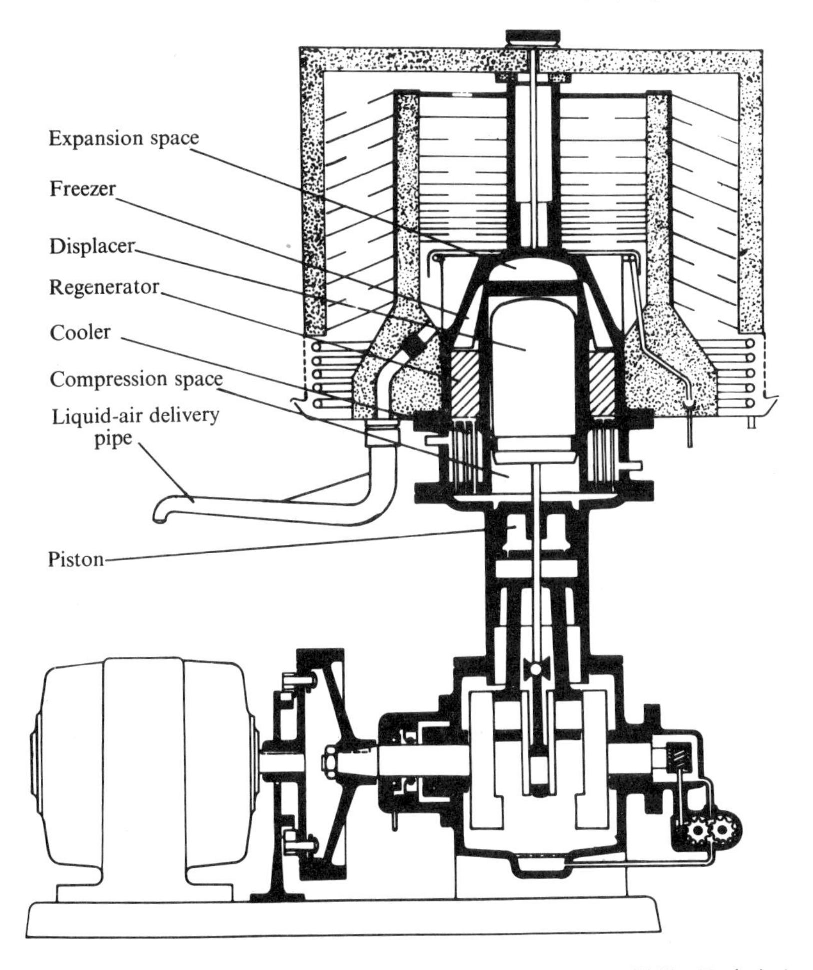

Fig. 8.10. Type A Stirling-cycle cryogenic cooling engine. (Courtesy Philips Technical Review.)

contact with the cooled head. In passage through the cooler, water and carbon dioxide was precipitated on the plates. The build-up of the solid material was sufficient to require (every 100 hours, or so) a 'defrosting' process, taking about two hours. Subsequently, an air-separation column was introduced, for use in combination with the air liquefier, as shown in Fig. 8.11. The output in this case is high-purity nitrogen, at the rate of five to six litres per hour. Four-cylinder versions of the liquefier were developed, also, along with an appropriate air-separation column. The output of these machines is about 4000 litres per week of liquid nitrogen. They can operate virtually unattended for up to two weeks between defrostings.

Another extension of the basic liquefier involved the addition of a second expansion space. With this machine, it was possible to attain a minimum temperature of 12K. Refrigeration was obtainable at two temperatures, simultaneously, in the range from 12 to 40K, in the cold exchanger, and 50 to 80K, in the intermediate exchanger. This machine provides a basic temperature-control system for cryogenic research work, and has been incorporated, also, by Philips into a hydrogen–neon liquefier and recondenser. It has been used, also, as part

Fig. 8.11. Type A cooling engine with air-separation column. (Courtesy Philips Technical Review.)

of a helium-liquefaction system, including a Joule–Thompson expansion unit.

Cooling engines very much smaller (and very much larger) than the original liquefier have been marketed. An example of the small machines is shown in Fig. 8.12. These are miniaturized cryogenic cooling engines, designed for spot-cooling of infra-red detectors and other solid-state electronic devices (as well as superconducting materials), cryo-pumping, and recondensing cryogenic fluids. Machines are available to provide temperatures down to 30K in a single-stage

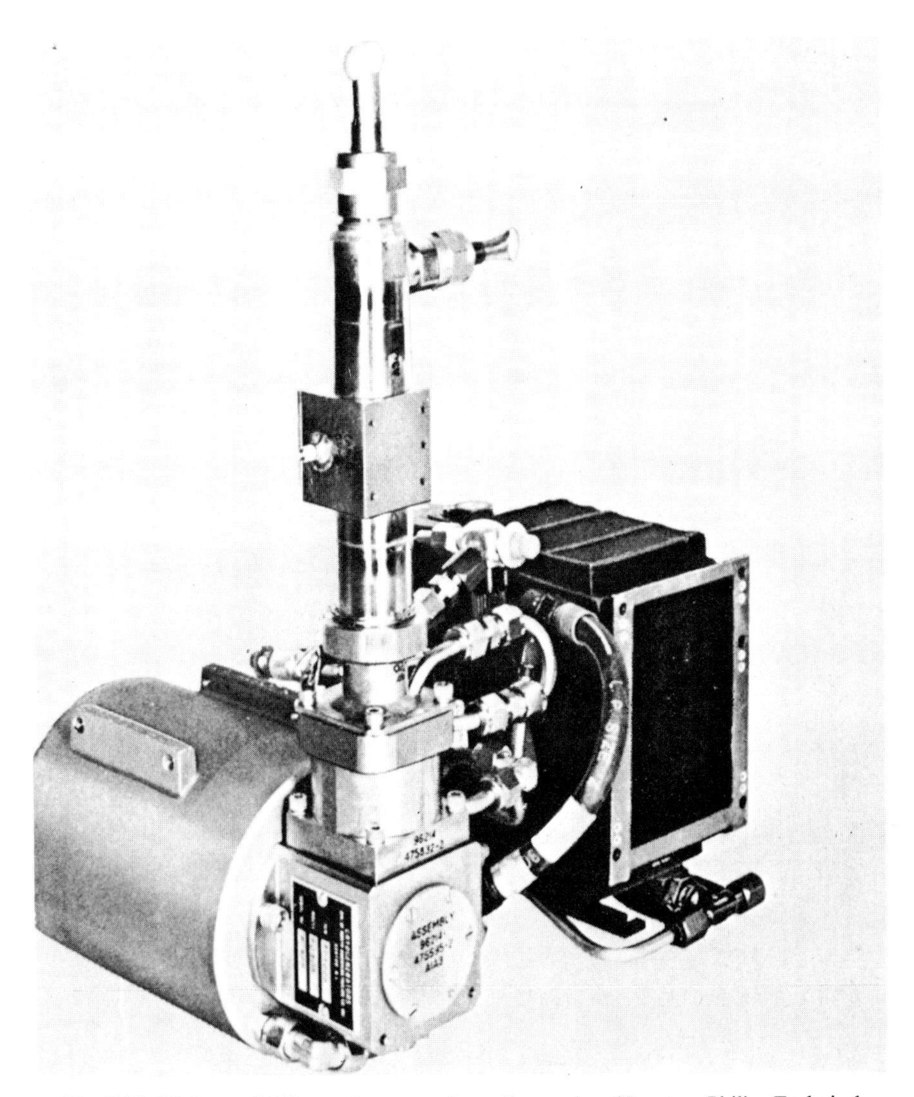

Fig. 8.12. Miniature Stirling-cycle cryogenic cooling engine. (Courtesy Philips Technical Review.)

expansion, and to 20K in double-expansion versions. A triple-expansion version has been evaluated and temperatures as low as 7K attained. A very small machine, weighing about 1 kg complete, has been developed (Daniels and du Pre 1971). At the other end of the scale is the large industrial refrigerating engine, described by Dros (1965), and shown in Figs. 8.13 and 8.14. The performance of this engine is shown in Fig. 8.15. This machine is of particular interest, since it is the only Philips machine that is not of the single-cylinder piston–displacer variety. Instead, the large cooling engine is of the two-piston type, with the pistons driven hydraulically: this has resulted in significant gains in 'efficiency' compared with piston–displacer machines. The large machine is designed for use as an industrial unit, rather than for laboratory use, and operates with high reliability for long periods, without maintenance. It is expected to find considerable application in the L.N.G. (liquid natural gas) field, for recondensing the L.N.G. boil-off from large storage vessels and marine transports.

Köhler and his co-workers stress the importance of the regenerator in the operation of refrigerating engines. The regenerator is much more important in the case of the low-temperature cooling engine than it is to the prime mover. Kohler points out that a 1 per cent inefficiency in the regenerator induces a loss of 21 per cent of the ideal cold production, at an expansion temperature

Fig. 8.13. Large industrial Stirling-cycle cryogenic cooling engine. (Courtesy Philips Technical Review.)

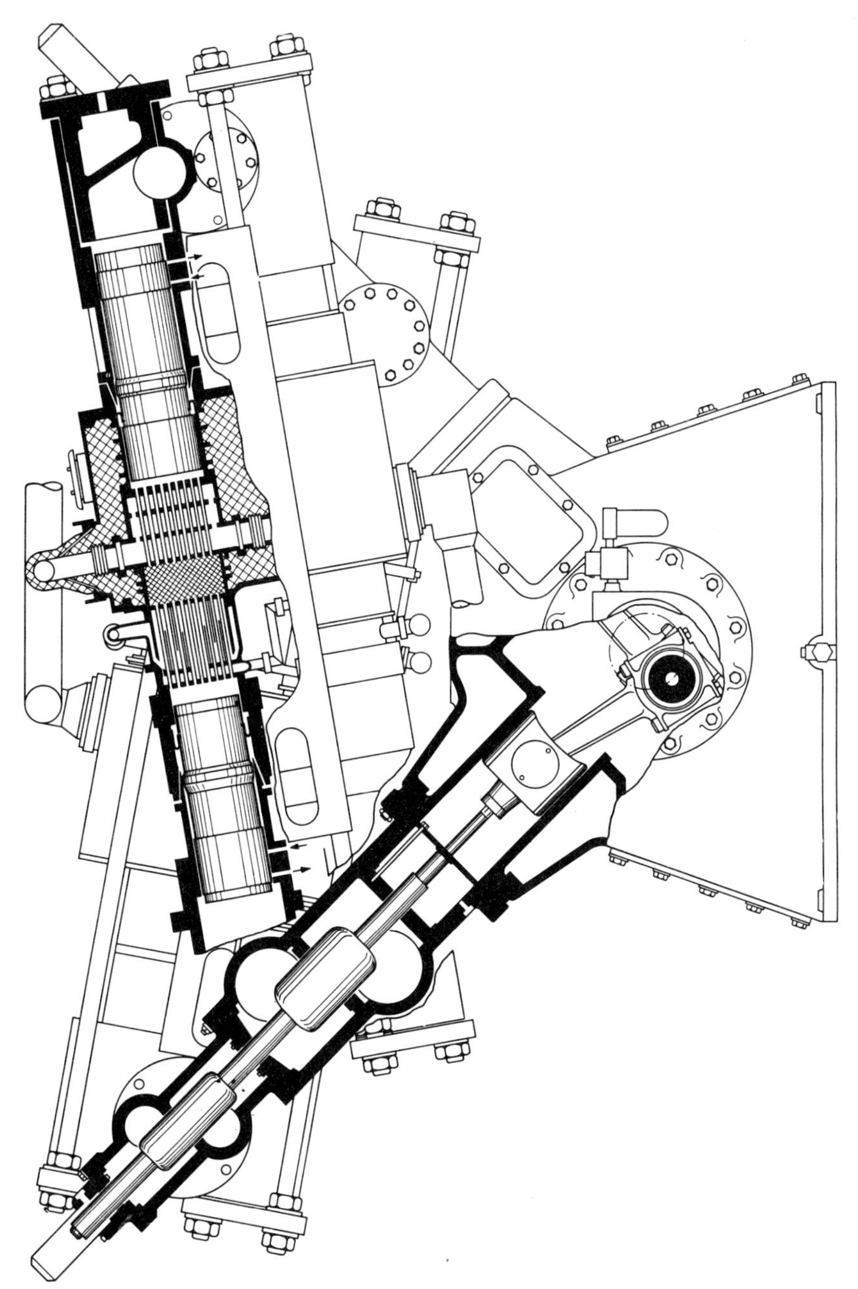

Fig. 8.14. Cross-section of a large cryogenic cooling engine. (Courtesy Philips Technical

of 75K, which increases to 98 per cent of the ideal cold production, at an expansion temperature of 20K.

In their development of a large industrial cooling engine, Philips have designed a machine utilizing rhombic drive. This was later abandoned in favour of the Dros hydraulic two-piston machine, described above. Development of the rhombic-drive engine was continued by the Werkspoor Company, and some machines were marketed in the late 1960s by the C.V.I. Corporation. It is not known how many were made, or sold, nor whether the machine is still in production.

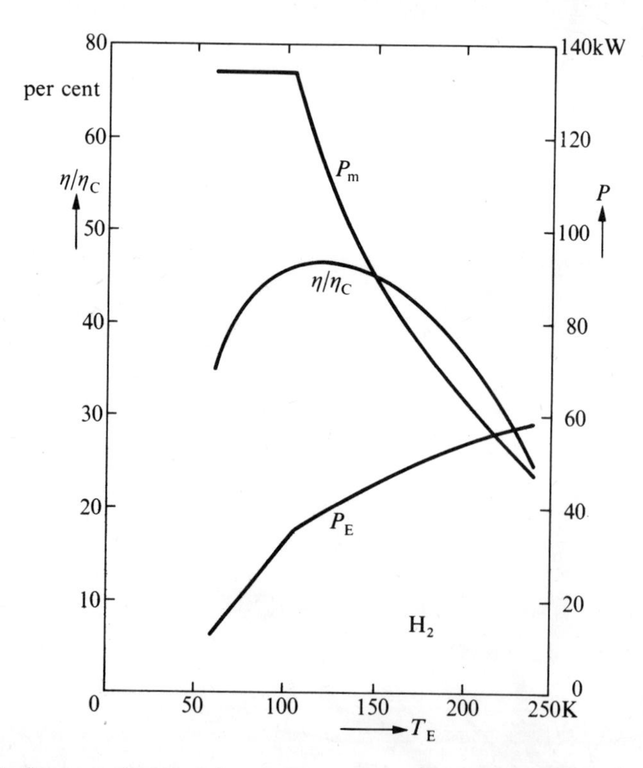

Fig. 8.15. Performance characteristics of a large cryogenic cooling engine. (Courtesy Philips Technical Review.)

Licensees of Philips engines

General Motors became a licensee of Philips in 1958, and carried on various programmes of Stirling-engine research and development, until 1970 (Heffner, 1965; Paste and Whitaker 1961; Agarwal, Mooney, and Toepel 1969; Lienesch and Wade 1969). More recent licensees include the 'Entwicklungsgruppe Stirling Motor M.A.N.–M.W.M., Stuttgart, in 1967, United Stirling A.B. of Malmo,

Sweden, 1968 (Neelen, Ortegren, Kuhlmann and Zacharias, 1971), and the Ford Motor Company, Detroit, in 1972.

All these licence arrangements were for prime movers. So far as is known, there are no special licence arrangements for cooling engines. These are manufactured and distributed by Philips.

9 Applications of Stirling engines

IN order to define potential legitimate applications of the Stirling engine it is necessary first to review both the advantages and disadvantages of the machine.

Advantages

The advantageous characteristics of Stirling engines may be summarized as follows: Stirling engines run without noise or vibration; burn anything with low rates of air pollution; can utilize a heat supply from any source including stored thermal energy from an intermittent electrical (or combustion) supply, concentrated solar energy, or radioisotope and nuclear-reactor heat. In their advanced form they can be superior to petrol engines in specific output, and can match a diesel engine in part-load economy. They have a maximum thermal efficiency better than any thermal-energy conversion device. They have an excellent part-load torque characteristic, and the cyclic torque variation is better than any other reciprocating engine except, perhaps, a double-acting steam engine. The combustion products are not in contact with the moving parts, so that wear is minimal and, consequently, long life and low maintenance are possible. The consumption of lubricant oil (an expensive commodity) can be virtually eliminated, and its renewal-interval greatly extended.

Stirling engines exist in a wide variety of forms, some of which are exceedingly simple, and can be made in sizes from single-cylinder toys, rated at fractions of a watt, to seemingly-unlimited upper sizes of thousands of horsepower per cylinder. They can be used as prime movers, cooling engines, and heat pumps. As cooling engines, at cryogenic temperatures, they are superior in size and performance to anything other than their open-cycle cousins.

Disadvantages

The principal disadvantage of Stirling engines is that engines of advanced form are both costly and complex. Simpler versions are less costly, but do not have the efficiency and specific output of internal-combustion engines. At the present time, it appears unlikely that the cost per horsepower of advanced Stirling engines could be reduced to that of a diesel engine, even discounting cost-differences which depend on the numbers of engines in production. Stirling engines of high thermal efficiency and high specific output also pose formidable problems in heat-transfer methods and seal-technology, both in terms of materials and design. The following discussion is a summary of the principal difficulties.

Heating

The upper-temperature heat-exchanger is, perhaps, the major problem area. Parts of this unit must operate continuously at the maximum cycle-temperature, and must be constructed of special high-temperature alloy or ceramic materials to gain high thermal efficiency. At the same time, the internal void-volume, contributing to the dead space of the engine, must be kept to a minimum. If the volume is small, the area for heat transfer will be small, so that very high heat-fluxes must be sustained. One approach to solving the problem is to use a heat-pipe (Chap. 8), which acts as a thermal transformer. In order to gain a high specific output, the engine will be pressurized to a high degree, using hydrogen or helium, not air. However, many materials at high temperature and pressure tend to be slightly porous to these gases. The upper-temperature heat-exchanger thus includes a unit which operates continuously at temperatures close to the metallurgical limit, with alternating high internal pressure and pulsating flow of hydrogen or helium. This embraces the frontiers of technology in the fields of materials, joining and fabricating techniques, and thermo-fluid design.

The upper-temperature heat-exchanger also includes a secondary heat-exchange unit, for preheating incoming air from the combustion products. This is a most necessary process where high efficiency is required to minimize the 'exhaust-stack loss'. The process must be accomplished at essentially atmospheric pressure, so that the density of the heat-transfer fluids is relatively low. This implies the need for a large surface area for heat transfer, to gain the high exchanger-effectiveness required. The requirement for high specific output imposes limits on the size and weight of the heat exchanger, so that ingenious design is required. It may be of either the recuperative, or the regenerative, type.

Cooling

Another severe problem in high-duty Stirling engines is cooling in the low-temperature heat-exchanger.

Of the heat supplied at high temperature to the engine, some is immediately lost to the exhaust or stack. This is a total loss, and contributes nothing to the performance of the engine. Of the remainder actually transferred to the engine, some is converted to work, and the rest is rejected (as 'wasted heat') to the cooling system.

In the case of the diesel or petrol engine the 'wasted heat' is split between that handled by the cooling system, and that blown up the exhaust stack. Furthermore, in the internal-combustion engine, an increase in coolant temperature improves thermal efficiency. The reverse is true in the case of the Stirling engine, since any increase in the minimum cycle-temperature of the working fluid adversely affects both the specific output and the thermal efficiency. Thus, compared with a diesel engine, the cooling system of a Stirling engine must not only handle more heat per horsepower, but must do so more effectively, in order to keep the coolant temperature as close to atmospheric ambient as possible.

The radiator for a Stirling engine might be two or three times the size of that for a diesel engine. Prospects for a high-performance Stirling engine, with simple air-cooling and no secondary fluid loop, are virtually zero.

Seals

Sealing is another major problem of high-performance Stirling engines. It is important because the fluid involved is not air (which could be readily replenished). Leakage, however slight, cannot be tolerated over the long term, unless the weight of highly-compressed-gas bottle storage is included in the engine inventory.

Other problems that will engage the interest of the engine designer include thermal stresses involved in the transition from a high-temperature to low-temperature region in the regenerative zone, the avoidance of 'hot spots' in the high-temperature exchanger, fatigue failure due to pulsating stress, engine balancing, governing, and control, and the usual problems of bearings and lubrication.

Prospective applications

Automotive engines

The modern advanced Stirling engine is the product of a combination of multi-disciplinary technologies. The Philips Research Laboratory is possibly the only establishment in the world with the proper combination of advanced technology and far-sighted management, under which long-term large-scale research can be carried out. Despite the seemingly prodigious total of nearly 2000 man-years of effort already spent on Stirling-engine research, it is trivial compared to that invested in internal-combustion engines. Every year, General Motors, alone, has thousands of engineers working on internal-combustion engines. The sum-total of human effort expended on internal-combustion engines, from university professors at their research – through design, production, and maintainance – to the newest apprentice garage hand, is an incalculable figure. That Philips, in a relatively short time and with a relatively small team, were able to produce engines superior in performance to all others is incredible, and with this in mind, it is hazardous to try to foresee future developments.

It is doubtful that the Stirling engine, on strictly economic grounds, could ever offer a challenge to the diesel engine in any established sphere. However, affairs are no longer decided on straightforward techno-economic considerations. The social aspects of engineering are becoming more important; this is particularly noticeable in the interest which the public now takes in air pollution by automotive engines. It is fitting that the interest is highest in the United States, where the automotive horsepower-capacity is greatest. At the present time (1971), both Federal and State legislation is proposed to severely curtail exhaust-emission from automotive engines. At the same time, Philips have already convincingly demonstrated the significant reductions in both noise and air pollution that would follow from the adoption of the Stirling engine for automotive use. It may be that social pressures will force the adoption of severe standards for controlling air pollution, and both diesel and petrol engines will require such derating (or complicated emission-control systems) that the Philips engines will become competitive. The chance of this situation developing is a real possibility, and is, without doubt, the reason for the present emphasis, by Philips and their licensees, on automotive engines of 100 and 200 horsepower. The next few years will be critical, but, by 1980, Stirling engines in trucks may be no longer a novelty.

Cryogenic cooling engines

Philips interest in Stirling-cycle cooling engines started as virtually a chance by-product of their early work on prime movers. The Philips family of cooling engines was developed under the direction of Dr. J. W. L. Köhler, and now occupies a commanding place in any inventory of small- to medium-scale cryogenic cooling systems. The cooling engines have led to the development of associated cryogenic engineering equipment. Indeed, it has been suggested that the profits of the Cryogenic Division have, in fact, paid Philips all the costs expended on the prime mover research. The types of cryogenic cooling engines available commercially from Philips cover a wide range, from miniature models, having a cooling capacity of fractions of a watt, to the large industrial machine, rated in kilowatts of cooling capacity. At present, the market is fairly limited, and is well covered by existing machines.

With the development of higher-temperature superconducting materials, and the increasing application of infra-red techniques, there appear to be manifold opportunities for small cooling engines. Machines that are smaller, cheaper, and more reliable than those presently available will be required.

In addition to North American Philips, other companies actively marketing small cooling engines include Malaker Laboratories, N.J., and the Hughes Aircraft Company, California. Several other companies also have research interests in this field.

Refrigerating machines

Stirling engines can be made to provide refrigeration at any temperature level, in the range from atmospheric ambient to near absolute zero. They are well established for use as cryogenic cooling engines, at temperatures less than 100K. At higher temperatures, they are virtually unknown, although there is no intrinsic technical reason why they should not be used as refrigerating machines at temperatures near atmospheric.

A comparison was made, by Finkelstein and Polanski (1959), of the performance of a two-cylinder Stirling-cycle engine and a standard vapour-compression refrigerating machine. They found the coefficient of performance of the Stirling-cycle machine to be less than the vapour-compression machine, at temperatures greater than 230 K, and to be increasingly superior below this temperature. The Stirling-cycle machine was made on a shoestring budget, with no attempt to optimize either the performance or design. Very little experimental work on relatively-high-temperature refrigerating machine appears to have been carried out.

The author has investigated (theoretically) the possibility of Stirling-cycle vehicle-cooling systems. With a Rinia-cycle unit, having a swashplate drive (as shown in Fig. 9.1), it was possible to anticipate considerable savings in both weight and volume, with no loss of performance compared with vapour-compression machines. There appear to be promising applications for this type of unit in railway coach (and bus) air-conditioning systems, and in the cooling systems of military vehicles and naval craft.

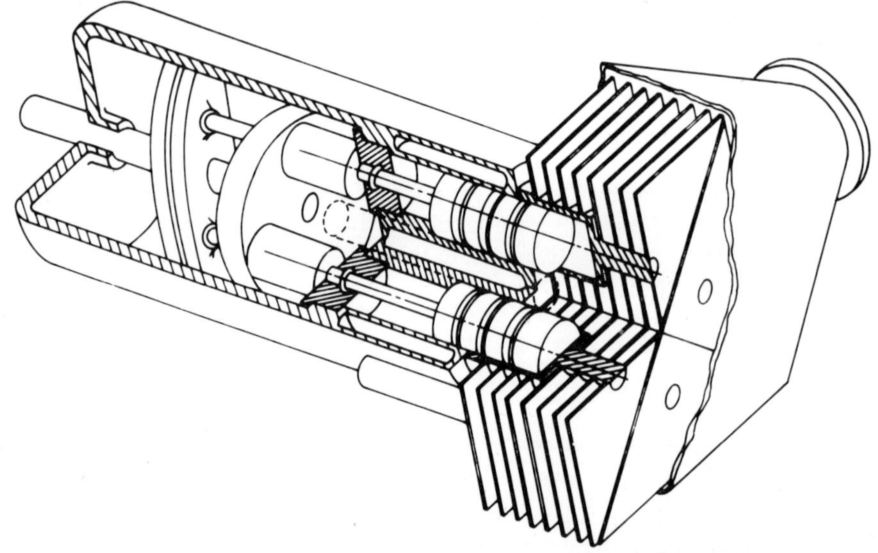

Fig. 9.1. Vehicle-cooling unit. The figure is a representation of a four-cylinder Rinia-arrangement Stirling-cycle cooling engine, prepared in a design study for engine-driven vehicle-cooling units. The study indicated the possibility of appreciable savings in weight and size for Stirling-cycle units, compared with vapour-compression machines. There are possibilities for the use of units of this kind in automotive, railway, and marine air-conditioning.

Another possibility for air-conditioning units is the duplex machine shown in Fig. 9.2. This is a combination of thermally-activated prime mover, producing work to drive a cooling engine mounted on a common crankcase. It might be used as a gas- or oil-fired air-conditioning unit, as an alternative to the thermally-activated absorption-refrigeration machine. Preliminary studies have shown that appreciable gains in efficiency and great savings in size and cost might be achieved. The advent of a gas- or oil-fired air-conditioning unit, of reasonable size, cost, and reliability, would have a significant effect on the energy economy of the United States.

Small electric-power generators

Many applications exist for small electric-power generators, capable of operating unattended for long periods and in remote locations. The significant power-level varies from five watts to five kilowatts, with a concentration of interest in the 200 to 500 watt level. The generators are required for a variety of purposes, but, principally, to supply power to navigation aids in lighthouses or buoys, to automatic meteorological stations, or to telemetry and communications booster-stations. There are both civil and military applications underwater, in mountainous, inaccessible regions in the Arctic, and on dangerous navigational hazards. In most of these applications reliability is the required criterion. There are virtually no restrictions on size, weight, and speed. Neither is starting and stopping important, since, in most cases, it is possible to include an electric storage-battery system, to allow power to be drained off at a high rate (at night, for example), whilst the engine generates continuous power at a steady rate. Engine regulation and control is not an important matter. In most applications, it is preferable (and easier) to adjust the system electrically, rather than control the engine.

Thermal efficiency is an important factor, it determines the necessary radio-isotope, or fossil-fuel, inventory for heat supply. However, an over-all thermal to electric conversion rate of 20 per cent is considered excellent, since the only

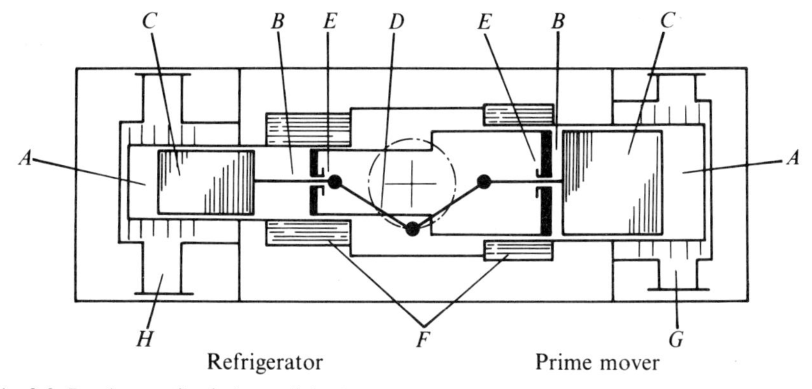

Fig. 9.2. Duplex gas-fired air-conditioning unit.
A – expansion space, *B* – compression space, *C* – regenerative displacer, *D* – drive mechanism, *E* – piston, *F* – cooling system (heat rejected from the unit at ambient temperature), *G* – heater (heat supplied at high temeprature, by combustion), *H* – freezer (cools air passing over the expansion space).

† In this unit, heat is supplied at high temperature to a mechanism acting as a Stirling-engine prime mover, converting some of the supplied heat to shaft-work, and rejecting heat at ambient temperature from the compression space. The shaft-work produced is consumed in operating a second Stirling-cycle unit, acting as a cooling engine. This unit rejects heat, at ambient temperature from the compression space, and absorbs heat, at a low temperature from the surrounds of the expansion space. The heat absorbed at low temperature is the useful refrigerating product of the duplex unit. Studies indicate that machines of this kind would be appreciably more efficient than thermally-activated absorption refrigerating machines, with great savings in size and weight.

other available systems are thermoelectric units with conversion ratios of 8 per cent, or less. Neither diesel engines nor petrol engines with sufficient reliability are available in the size range under consideration. Stirling engines appear ideally suited for this duty, and it is ironic that Philips abandoned their development of the small air-cooled electric-power generator, which was a good approximation to the above requirement. Given a few additional years of research, this unit might have proved to be a substantial success in both civil and military, maritime and land-based, applications.

Research leading to the development of a small radioisotope-fuelled Stirling-cycle engine has been undertaken by the author at the University of Calgary, at the request of the Atomic Energy of Canada Ltd. (1969). It was intended that the system utilize Cobalt 60 as the fuel, to be converted to thermal energy resulting from the radioactive decay of the material, at an over-all conversion ratio of 20 per cent. The half-life of the Cobalt 60 isotope is 5 years, and an operating

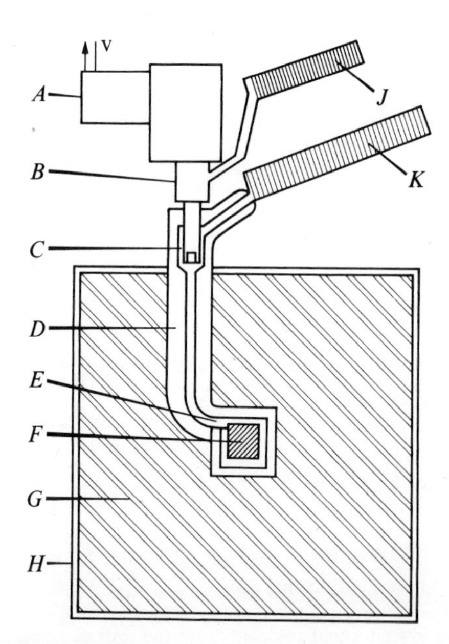

Fig. 9.3 Isotope power system.
A – electric generator, *B* – Stirling engine, *C* – heat-pipe condensing section, *D* – thermal insulation, *E* – heat-pipe evaporative section, *F* – energy source, matrix of isotope fuel, and absorber, *G* – biological shield, *H* – fire and impact shield, *J* – cooling-system heat-exchanger (condensing freon heat-pipe), *K* – temperature control and thermal dump heat-exchanger (gaseous section of two-phase two-component heat-pipe).†

† The figure shows a conceptual arrangement for an isotope-fuelled electric-power system, for long-term unattended power supplies, in remote areas, for weather study, navigation, communication, and defence purposes. A Stirling engine is used to convert heat, produced by the radioactive decay of the isotope, to mechanical work, for driving the electric-power generator.

life of $2\frac{1}{2}$ years for the generator was anticipated. A block diagram of the proposed system is shown in Fig. 9.3. The energy source is a series of Cobalt 60 pellets, about $\frac{1}{4}$ inch in diameter and $\frac{1}{4}$ inch long, contained in a matrix of dense material, such as tungsten or uranium, and encapsulated in a suitable metallic sheath. Gamma radiation, emitted by the Cobalt 60, is decellerated by the matrix, and the energy of the radiation is dissipated in heating the matrix. An operating temperature of 600 °C is possible. Not all the gamma radiation is retained by the matrix, and a heavy biological shield of lead, some 2000 kg in weight, is required, with additional external fire and safety shielding of mild steel plate. The heat generated within the matrix is absorbed in boiling the liquid-metal component of a heat-pipe system, coupling the energy source to the Stirling engine which drives an electric generator and is located outside the lead shield. It would be possible to locate the engine and generator adjacent to the energy source, but this would increase the required mass of lead shield to a level beyond the payload of all but the largest helicopters. At least one 90° bend in the heat-pipe is necessary to prevent a radioactive 'shine' of gamma radiation along the pipe.

Use of the heat-pipe is advantageous, because it allows a ready mechanism for 'dumping' excess thermal energy. An automatic dissipative thermal dump is necessary as a safety precaution in the event of engine seizure (or other stoppage) to prevent overheating (and eventual vapourization) of the energy source, with consequent release of radioactive products. Further, the thermal output is initially greater than the design-load by an amount sufficient to compensate for the progressive decrease in thermal release resulting from the finite half-life of the isotope. For Cobalt 60, with a five-year half-life, the thermal output is down by approximately 25 per cent at the end of $2\frac{1}{2}$ years.

The proposed heat-pipe is a two-component two-phase system, consisting of a gaseous component of low thermal conductivity and a two-phase liquid-and-vapour metallic component of high conductivity. Normally the two components are separated quite distinctly by gravitational forces. The heavier conducting metallic component is contained in a well-insulated main section of the pipe, connecting both the energy source and the engine. The lighter low-conducting gaseous component is contained in an upper radiative section.

At design temperature, the interface of the metallic and gaseous components is within the insulated section, but, when the thermal input exceeds the design condition, the temperature rises, causing an increase in pressure. The interface then ascends the heat-pipe as the gaseous component is compressed, so that the high-conductivity vapour enters the radiative component, and heat is released, until the interface is restored to the insulated section. The proposed cooling system includes a second heat-pipe, using one of the 'Freon' refrigerant fluids which boils on the cooler section walls and condenses in the air-cooled radiative section. This is an alternative to a simple cold-water circuit which includes a pump driven by the engine crank.

Good progress has been made in the development of a suitable engine for the system. Fig. 9.4 is a drawing of a prototype unit, which was made in 1971 at the University of Calgary and embodied the results and experience of an earlier research engine. Preliminary studies are also in progress, sponsored by the Trinity House Lighthouse Service, directed to the possibility of operating this engine on combustion heat. There is a need for an engine, fuelled by diesel oil or liquid petroleum gas, to operate unattended for a period of up to one year on navigation buoys and lighthouse beacons. The Chipewyan unit is intended to be the prototype of a simple machine, available as a single (or V) engine configuration, using air as the working fluid and with a plain diaphragm compressor

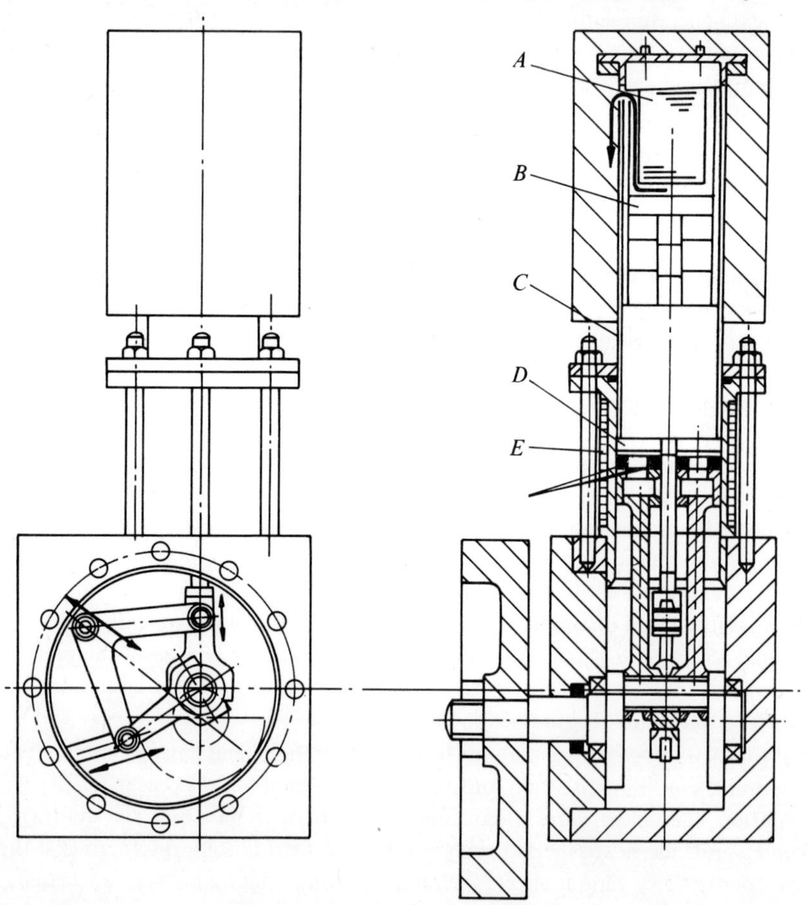

Fig. 9.4. Experimental 'Chipewyan' Stirling engine. Bore, $2\frac{3}{8}$ in, stroke, $1\frac{1}{4}$; pressure, 250 lb in^{-2} fluid air; 600 rev./min; rating, 200 watts shaft-work.
A – electric heater, *B* – expansion space, *C* – regenerative annulus, *D* – compression space, *E* – cooling jacket, *F* – Rulon seals.

This unit is under development at the University of Calgary, as part of a programme sponsored by the Atomic Energy of Canada Limited, Commercial Products, for small isotope-powered systems.

(operated by the engine) to maintain pressure in the crankcase. It will be a rather heavy slow-running engine of moderate efficiency, with emphasis on reliability, for long-term unattended operation.

Marine engines

The use of Stirling engines for marine purposes is attractive, because readily-available coolant supply resolves one of the principal difficulties in automotive Stirling engines. This was recognized by Philips at an early stage, and one of their first practical applications for the engine was in a motor cruiser (the 'Johann de Witt'), used to demonstrate the quietness of the engine.

Elsewhere, the Electromotive Divison of General Motors at La Grange, Illinois, worked on a V8 (800 horsepower) engine, intended for coastal vessels such as tugs which require manoeuverability of a high order. This engine included sparate shafts for the pistons and displacers. The shafts were coupled by a planetary gear, and, by manipulation of this gear, the phase angle 'α' could be changed. This provided the means to readily vary the output of the engine, and means to reverse the engine. In 1966, the author saw one bank of four cylinders mounted on the crankcase of the machine. So far as is known, no details of the engine were ever released, and the programme is now thought to be at an end.

The future use of Philips-type Stirling engines, of several hundred horsepower, for both military and civil purposes, appears likely, whenever the diesel engines that are now used are considered unsuitable on grounds of noise, vibration, or air pollution. In very large sizes (thousands of horsepower), it is likely that Stirling engines could be developed with advantageous characteristics, in terms of efficiency and specific output, compared with modern diesel marine engines. However, the development of such engines is unlikely, because of high development costs and the capacity of diesel-engine builders to satisfy the present limited market.

At the other end of the spectrum, there does appear to be the need for a small low-horsepower (up to five or ten h.p.) engine, for electric power and auxiliary-propulsion purposes, on private yachts and other luxury craft. For this application, quiet operation, reliability, and low maintenance, perhaps, would be the preferred characteristics, with cost per horsepower, efficiency, and specific output being items of lesser significance.

It is possible that a very satisfactory oil- or l.p.g.-burning unit, of three or four cylinders, could be developed, using air, at up to 10 atm pressure, as the working fluid, and running very smoothly, relatively slowly, and very quietly. Such a machine would be attractive to many boat owners.

Underwater power systems

The Stirling engine appears well suited for inclusion in a variety of underwater power systems, where electric work or mechanical power might be required on either an intermittent, or continuous, basis.

With radioisotope heat, the Stirling engine has the advantage of high efficiency over the competing thermoelectric system. With non-radioisotope heat, the Stirling engine benefits from its non-critical acceptance of virtually any thermal energy (either storage or combustion system), and its ability to operate silently with no valves and no periodic explosions. In combustion systems, fuels can be used that combine to produce reaction products that condense for 'on-board' storage so that no release of exhaust products is necessary. This permits the construction of closed systems, able to operate at great depths, and without 'exhaust-trail'.

Considerable work on advanced Stirling-cycle engines, with thermal storage, for underwater applications, was done at General Motors. The experimental development of a four-cylinder engine for underwater power systems, using hydrogen peroxide as oxident, is in progress at United Stirling, Malmo.

Solar-powered engines

There appears to be a virtually unlimited market for small cheap solar-powered engines (for use in undeveloped tropical countries) to operate water pumps for irrigation, and to drive low-power electric generators, supplying batteries for night illumination. This is by no means a new application; Finkelstein (1959) illustrated an early solar-powered Stirling engine, built by Ericsson in the nineteenth century.

The principal difficulty in providing machines for this market appears to be the cost. In addition to the engine, it is necessary to provide a solar collector and concentrator. These are devices which absorb incident solar radiation over a large area, and concentrate it on a very small area. The energy flux at the focus is very high, so that the high temperatures which are necessary for operating heat engines are achieved. A variety of collector shapes are possible, perhaps the best-known being the simple paraboloid (the shape of a searchlight or car-head-lamp reflector). Another form of collector is the Fresnel lens, shown in Fig. 9.5. This consists of a flat sheet of transparent material with grooves cut into the top surface at different angles, so as to bend the incident light rays to a common focus. Such a lens can be made relatively cheaply by being pressed out like gramophone records. One, purchased by the author in 1970, was about $2\frac{1}{2} \times 3$ ft, and cost $20. Unfortunately, the lens was made of plastic, with a tendency to become opaque in strong sunlight.

The incident solar energy is greatest if the collector is normal to the axis connecting it and the sun. To maintain maximum output over the longest possible time, an orientation mechanism is required to move the collector in sympathy with the daily transit of the sun in the sky. The orientation mechanism can be driven by a simple clock-mechanism, with periodic adjustment of the latitude angle to account for seasonal variation of the solar transit. A Stirling engine could be used to convert the concentrated solar energy to mechanical work for driving a water pump or electric-power generator. The engine might be in a variety of forms, from a Beale-type free-piston-engine water pump (the working space pressurized and hermetically sealed, during construction) to a simple Heinrici engine, running on air at a low pressure. It must be a cheap, reliable device, suitable for production on a large scale.

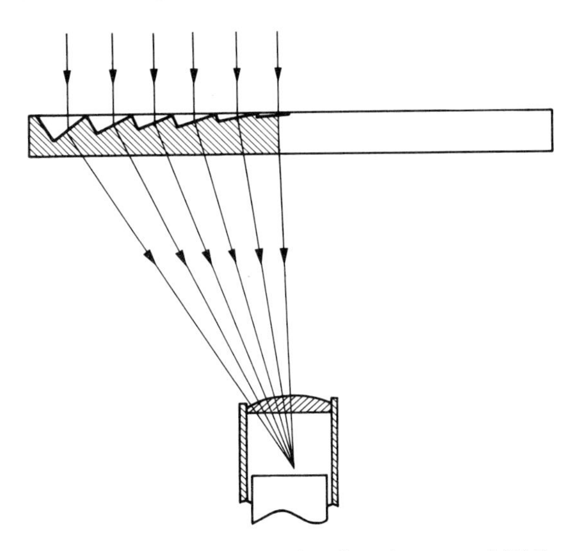

Fig. 9.5. Fresnel lens and quartz window for solar-powered Stirling engine.

One noteworthy feature is Finkelstein's invention of a quartz cylinder-head, so as to permit location of the engine cylinder with the focus of the collector within the cylinder. This provides the possibility for the working fluid to be heated to a very high temperature, with no particular requirement for high-quality materials in the hot-space enclosure.

Novelty items and teaching aids

In these days of computers, lasers, superconducting systems, and all the rest, it is surprising how a model Stirling-cycle engine still commands a high level of interest. People seem to enjoy looking at flywheels, cranks, and connecting rods, perhaps because it epitomizes, for them, childhood memories. At all events, there would seem to be a market for an 'executive toy'. This might be a small prime mover, utilizing combustion heat from a spirit lamp or visible electric-resistance element, from a battery pack or power point. Such a machine, if cheap enough for general distribution, could serve a useful educational function in acquainting the public with the possibilities of Stirling engines as quiet, low-pollution engines. The engine might also serve as the basis for development of a demonstration engine or teaching aid. The only Stirling engines known to be currently in commercial production fall into this category, ranging from the simple Bradley cylinder set (designed for incorporation into a 'Meccano' engine) and the Heinrici engine, designed by Sir Harry Ricardo and made by Cussons Ltd., to the more advanced Stirling arrangement of Leybold–Heraeus. Brief details of all these machines are given in Appendix 1.

Artificial hearts

During the 1960s, artificial hearts became a subject of engineering interest. They may be classified into two types: (a) devices which *temporarily* assume, and (b) devices which *totally* assume the duty of the normal heart.

The temporary devices are used, for example, in operating theatres, for maintaining blood flow during surgical operations on the heart. Another use is in providing the opportunity for a badly-affected heart to 'rest', during convalescence. In all cases, the temporary devices can be a large, heavy machine, external to the patient, and under the virtually continuous control of skilled personnel.

Machines of the other type, which totally replace the duty of the heart, do not yet exist. At present, the only way of replacing a human heart is to substitute another human heart, but there is some prospect that replacement by an animal heart or an artificial mechanical heart may be possible. Stirling engines might be used in future artificial hearts as the device to convert isotope heat to mechanical work.

Artificial hearts may be developed that use mains electricity (or a battery pack) to operate an electric motor, driving the blood-pump. This is not possible, at present, as there is no way to transfer the electric power from an external source through the skin. Furthermore, the need to remain 'plugged-in' to a power point, or a heavy battery pack, may impose unacceptable limitations on the mobility of the patient. The alternative is to provide a totally implanted device within the body, and fuelled with sufficient isotope to allow continuous operation for a suitable time (three to five years). The device would include an energy-conversion unit, to convert the heat, released by the isotope, to mechanical work. Some form of blood-pump, utilizing the mechanical work available, would also be included. The thermal–mechanical converter might be a Stirling-cycle engine.

A totally-implantable artificial heart is a concept that has many advantageous features. The present-day ethical problems of human-heart transplants would be avoided, and a stock of machines could be readily available, for use by surgeons at their discretion. However, a major problem, at present, is that there appears to be no alternative 'fuel' to a radioisotope. Many isotopes exist, but few of these would be suitable for an artificial heart. The choice would be limited to isotopes emitting alpha particles, which require minimum biological shielding. Other isotopes, emitting beta and, more particularly, gamma particles, require very heavy shielding. Unfortunately, the alpha isotopes are rare and expensive and, at present, no adequate source of supply exists. Even assuming that special effort was concentrated on the production of appropriate alpha isotopes, it is unlikely that the price could be reduced to a level within the reach of a reasonable number of people. Another difficulty is raised by the possibility that the recipient may meet an untimely death. He may be involved in a hotel, or car, fire, or he may die from natural causes, and be cremated before the isotope could be removed. In any of these cases, radioactive vapours may be released. Thus, thermal shielding, sufficient to survive cremation, may alone render the radioisotope artificial heart impracticable.

Not withstanding all these factors, exploratory research has been financed in the United States, by the National Heart Institute, a Government Agency. The programme was broadly based, and encompassed many facets of advanced energy-conversion and materials technology. The programme included two separate efforts on 'Stirling engines' which have been reported, independently, by Buck (1968) and Martini (1968). According to the definition which was adopted earlier, neither of these engines were, in fact, Stirling engines. Rather, they were Ericsson engines of the fixed-working-volume displacer type, with gas-operated valves. The principal difference between them was that the Buck engine had a fixed, separate regenerator and the Martini engine had a regenerative displacer. The operation of nuclear-powered artificial hearts has been described by Harmison *et al.* (1972), including development to the point of installation of an isotope-fuelled artificial heart in a calf, which was fully-awake and free to move about.

The conversion of thermal to mechanical energy, with absolute reliability and reasonable efficiency, is perhaps the real nub of the engineering problem of the artificial heart, and, from every point of view, radioisotope-fuelled Stirling engines remain as strong contenders in any consideration of the converter unit. One possible line of approach is shown, diagrammatically, in Fig. 9.6. In this

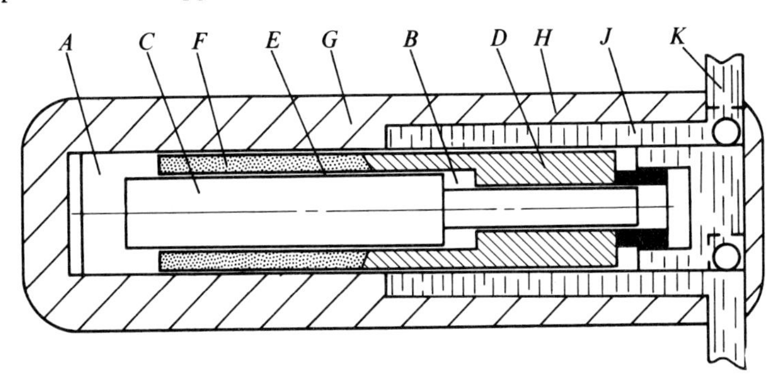

Fig. 9.6. Conceptual diagram of Beale free-piston Stirling-engine heart-pump unit. A – expansion space, B – compression space, C – displacer, D – piston, E – regenerative annulus, F – porous isotope matrix, G – thermal insulation and radiation shield, H – bypass cooling system, J – cylinder casing forming pump ram, K – high pressure saline solution.

In this concept, the heat released by radioactive decay of the isotope fuel is converted to work in a (relatively) high-frequency Beale free-piston Stirling engine. The work available is consumed in pumping saline fluid, used to operate a low-frequency blood-pump. The Beale engine is self-starting, and requires no external gas seals (see Chap. 10).

arrangement, a modified Beale-type free-piston Stirling engine operates a simple plunger-type hydraulic pump. The radioisotope fuel is attached to the heavy piston, and encloses the displacer, in the form of an internally-heating porous annular heat-exchanger. The displacer and the case are lightweight compared with the piston and isotope heating element. In operation, the piston remains virtually stationary in space, and the outer case and displacer oscillate periodically to provide the necessary variations in the expansion and compression space. The frequency of oscillation would have to be relatively high (perhaps 1000 c/min) to yield a reasonable level of specific output. An extension of the case constitutes the ram of the hydraulic pump. With this ingenious arrangement, no external seals whatever are required. The unit can be pressurized and hermetically sealed during production. Any radon gas generated during the radioactive decay of the isotope will also be retained within the machine case. The unit is mounted on guides, and insulated as shown. The pump is supplied with an appropriate liquid, possibly a saline solution, which might also be used for cooling. The saline solution, at relatively high pressure, passes from the high-frequency engine-driven pump to a hydraulic-motor-driven low-pressure blood-pump. This must be a compound machine, of low frequency, and with some kind of flexing diaphragm for blood displacement.

Insufficient work has been done on this concept for the technical feasibility of reliable, efficient, long-term operation to be established. Nor is it possible, at this stage, to make meaningful estimates of the size, weight, and other critical characteristics of the system. The attractive features of the Beale engine in this application are

(1) that the engine is self-starting, a unique characteristic for single-cylinder engines,
(2) the arrangement is simple, direct, and does not require lubrication of bearings,
(3) gaseous seals can be dispensed with,
(4) the engine will operate in any orientation, so that the recipient may lie, stand, or sit, at will.†

Nuclear-reactor-type base-load electric generating stations

The concept of very large Stirling engines, perhaps of the Rinia variety, used as the thermal convertors in nuclear-reactor-type base-load power stations of multi-megawatt capacity, has been discussed, in general terms, by W. J. Bradley of the Atomic Energy Authority of Canada, Chalk River Nuclear Laboratories. Bradley proposes, to incorporate the cooling channel of the reactor as the heater of the engine, with helium, at high pressure, as the working fluid of the engine, and simultaneously acting as the coolant fluid of the reactor. The advantages of this system are manifest compared with other gas-cooled reactors

† For an excellent description of the function and operation of the heart see Longmore, D. (1971). *The Heart.* World University Library, Weidenfeld and Nicolson, London.

and secondary steam-raising heat-exchangers. The Stirling engine is simple, and less costly, because the engine requires no separate pumps or compressors for the primary coolant. The dead space in the engine must be minimal, thus forcing the deletion of ducts, feeders, etc. The reactor must be an integral part of the engine.

The reactor-heated Stirling engine seems an exciting possibility. Another possibility, with greater flexibility, is a modification to include valves, thus converting the engine to an Ericsson machine, but relaxing the stringent limits necessary on the 'dead space' in a Stirling machine.

Total-energy systems

Total-energy systems comprise a machinery ensemble which, supplied with fuel, air, and water, can provide the full range of utility demands, e.g. air-conditioning, electric power, and hot and cold water. Such installations, in office buildings, motels, hotels, residential buildings, department stores, or large shopping centres, typically require a prime mover in the 50 to 500 horsepower range.

The concept of total-energy systems was actively promoted in the United States during the 1960s, as one part of an effort to increase the commercial and domestic demand for gas in the summer months, and so equalize the load profile over the whole year. In the United States, air-conditioning (always taken to be cooling and dehumidifying) is as important a social amenity as heating. Gas is well suited for heating purposes, but less so for cooling. Serious efforts have been made to improve the performance, or reduce the cost, of gas-heated absorption-cycle refrigeration systems, but have, so far, been largely unsuccessful.

A number of total-energy systems have already been installed. In most systems, gas or fuel-oil is burned with air, and some form of prime mover is used to convert a fraction of the released chemical energy to mechanical work. The mechanical energy is used to drive an electric generator or the compressor of a vapour-compression refrigeration unit. Electric power may be generated at high frequency (400 c/s) for lighting purposes, and at relatively high voltages for power purposes, operating elevator motors, kitchen equipment, and refrigeration units. The exhaust-heat from the prime mover may be utilized for heating the building, raising low-pressure steam for kitchen and laundry, providing hot water for general purposes, or as the thermal source in an absorption-cycle refrigerating unit which produces chilled water (or brine) for air-conditioning.

It can be seen that a total-energy system is an attractive concept. Operations are no longer dependent on an external power supply, and there is the prospect of utilizing the energy efficiently. Unfortunately, there are manifold practical disadvantages; the machinery ensemble is expensive, reliability is not yet satisfactory, and maintenance costs may become high. In some large cities (in the United States) natural gas is no longer in abundant supply, and an oil reserve may be necessary, for use when the gas supply is interrupted. This increases costs, and requires a dual-fuel capability for the prime mover. Another problem arises because, in operation, it is difficult to match the electric-power loads to the heating and cooling profiles, so that efficient use of energy cannot be gained unless large reservoirs for low-grade thermal energy are provided. So far, there is little evidence that the manufacturers of prime movers, electrical gear, exhaust-heat boilers, and refrigeration plant are co-ordinating their efforts to produce a compact, complete package-unit. Total-energy systems tend to be designed by consulting engineers on an individual basis, putting together a number of items of equipment not especially well suited for the purpose. Costs are therefore high, and results not entirely satisfactory. Little real progress will be made until an effort has been made to produce reliable, well-integrated package-units, at reasonable cost.

At present, the prime movers used in total-energy systems are gas-turbine engines and reciprocating internal-combustion engines. Gas turbines have good reliability, but appalling efficiency, particularly under part-load operating conditions. The reciprocating gas engines have a better performance, but are not sufficiently reliable, so that maintenance costs are high. It would therefore appear that a quiet efficient reliable Stirling-cycle prime mover, with a multi-fuel capability, would find immediate application in total-energy systems. It is possible that the refrigerant loads could be met by a Stirling-cycle cooling engine, drive by the prime mover, and, possibly, the unit could include a multiple-cylinder engine, in which the cylinders act as a prime mover or cooling engine, depending on the local conditions. It is possible to envisage an electromechanical unit that could serve as a thermally-activated electric generator or an electrically-driven cooling engine, as required.

We must, however, be cautious in the application of the total-energy system. Hindsight suggests that total-energy systems have little justification in highly-developed communities which have cheap reliable electric-power supplies, but there does appear to be a place for total-energy systems in less-developed areas, such as remote motels, shooting and ski lodges, Arctic development camps, and desert oil installations.

10 Research topics

Beale free-piston engines

FREE-PISTON engines of the type invented by Professor Beale, of the University of Ohio, have many possibilities for future application. They are self-starting, a unique characteristic not shared by single-cylinder crank-controlled machines, and, further, they can be made in versions where no external gaseous seal is necessary. In this configuration, they can be pressurized and hermetically sealed, during construction, thereby providing for relatively high specific output and no possibility of contamination of the moving parts from outside dust. In this form, they are well suited for applications where maintenance might be a problem, e.g. for use in developing countries, for military applications, and in domestic appliances.

Beale engines perform in any orientation—vertical, horizontal, inclined, or upside down. They are amazingly simple in construction, and do not depend for their operation on springs, valves, or any kind of mechanical gadgets.

Cycle of operation

The Beale engine essentially consists of three components, a heavy piston, a light-weight displacer, and a cylinder, sealed at both ends. These three components are shown in Fig. 10.1. It can be seen that a displacer rod, of appreciable diameter, passes through the piston. The displacer rod is hollow and open, so that the interior of the displacer is coupled to (and is, in fact, part of) the space below the piston, called the 'bounce space'. The working space is that part of the cylinder which is above the piston, and is divided into the 'compression space', between the piston and the displacer, and the 'expansion space', above the displacer. There is a long thin annular space between the cylinder and the displacer, and this serves as the regenerative heat-exchanger between the hot expansion space and the cold compression space. A heater is provided for the expansion space, and a cooler for the compression space.

Consider the system initially at rest at the state 0, shown in Fig. 10.1. The pressure is the same in all spaces, and the temperature is ambient atmospheric, throughout. Now let the expansion space be heated by the heater. As the temperature increases the pressure p_w of fluid in the enclosed working space will increase from 0 to 1. As the pressure in the working space increases, it acts to move both the piston and the displacer down the cylinder. The force acting on

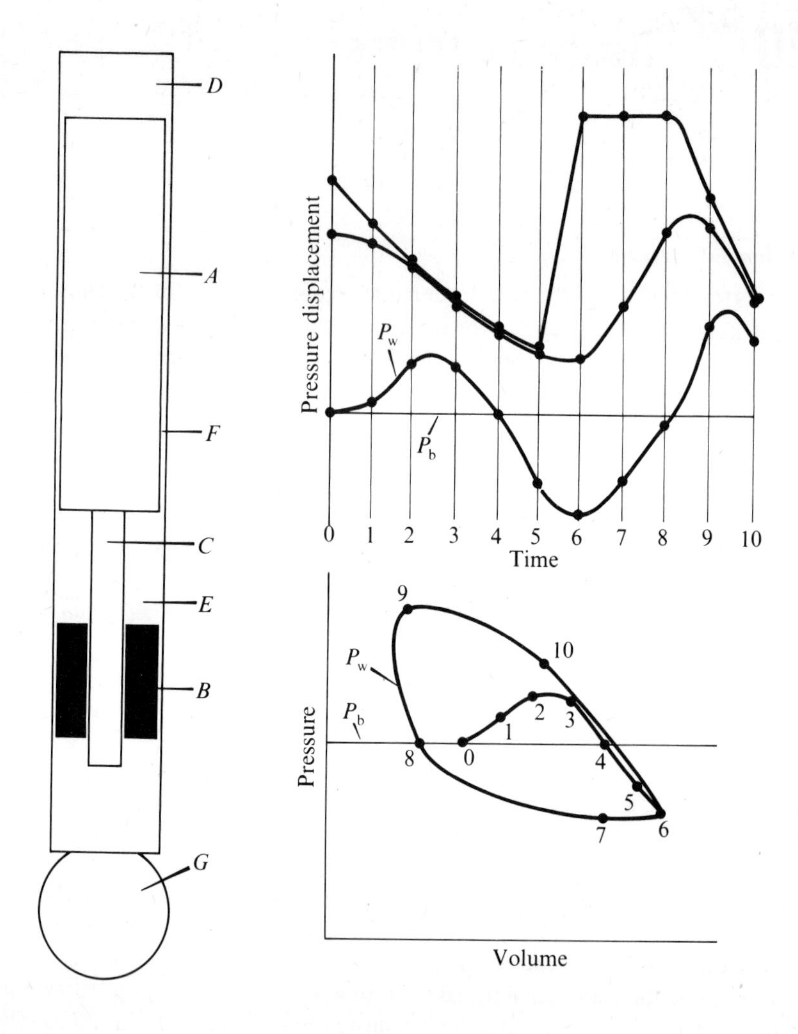

Fig. 10.1. Elements of a Beale free-piston Stirling engine.
A – displacer, *B* – piston, *C* – displacer rod, *D* – expansion space, *E* – compression space, *F* – regenerative annulus, *G* – bounce space.

p_w = pressure in the working space.

p_b = Pressure in the bounce space (assumed constant).

0 – Arbitrary starting condition.

1 – Heating of expansion space causes the working space pressure to rise above the bounce-space pressure. This causes the displacer and piston to descend, with the displacer accelerating faster than the piston, owing to the comparatively low mass of the displacer.

2 – the displacer contacts the piston, and both descend together.

4 – Working-space pressure is decreasing as expansion proceeds, and 4 is equal to the bounce-space pressure.

the piston is $(p_w - p_b)(A_C - A_R)$. The force acting on the displacer is $(A_R)(p_w - p_b)$. The acceleration of the piston downward is then

$$\alpha_P = (p_w - p_b)(A_c - A_R)/M_P$$

and the acceleration of the displacer is

$$\alpha_D = (p_w - p_b)\, A_R/M_D.$$

Now if the ratio M_P/M_D is large (i.e. 10:1) and if A_R/A_C is appreciable (i.e. $\frac{1}{4}$), then $\alpha_D > \alpha_P$. The displacer will accelerate rapidly therefore, along the cylinder, and cause the working fluid to leave the cold compression space, and enter the hot expansion space. This further accentuates the pressure rise above the bounce-space pressure (assumed to remain constant), and further accelerates the piston and displacer. Eventually, the displacer and piston contact (at 2), and, thereafter, move together. After the piston and displacer contact, there is no further flow of fluid into the expansion space, so that the pressure begins to fall as the expansion proceeds. At 3, the pressure p_w is still greater than p_b, so that the piston and the displacer continue to accelerate.

The expansion proceeds to the point 4, where the working fluid pressure p_w is the same as the bounce-space pressure p_b. The inertia of the heavy piston is sufficient to continue the expansion beyond pressure equilibrium, so that the pressure in the working space p_w now falls below the bounce-space pressure p_b, and, thus, decellerating forces (caused by pressure differences) are applied both to the piston and displacer. The displacer, being of light mass, is the first to respond. The decellerating force halts the downward motion of the displacer, and it separates from the piston, which continues down. Fluid then flows from the hot expansion space, along the length of the regenerative matrix, to the cold space. This causes a steep decline in the working-space pressure, so that a large pressure difference $(p_b - p_w)$ is established. The displacer is accelerated swiftly upwards to the top of the cylinder (at 6), and remains in this position so long as the bounce-space pressure exceeds the working-space pressure.

Eventually, at 7, the piston halts, and begins to ascend upwards, driven by the excess bounce-space pressure. As the compression proceeds pressure equilibrium is again momentarily established (at 8), and the working-space pressure rises above the bounce value. At this point the displacer begins to descend to contact the piston at 9, and repeat the cycle, but without the starting sequence 0–4.

---

5 – Working-space pressure is below the bounce-space pressure. The differential decellerates the piston, and at 5 is sufficient to cause the displacer to begin to rise up the cylinder. This displaces fluid from the hot expansion space, causing a further decrease in the working-space pressure, and accentuating the piston-decelleration and movement of the displacer.

7 – The downward motion of the piston is halted by the pressure difference between the working space and the bounce space, and, thereafter, the piston rises on the upward stroke.

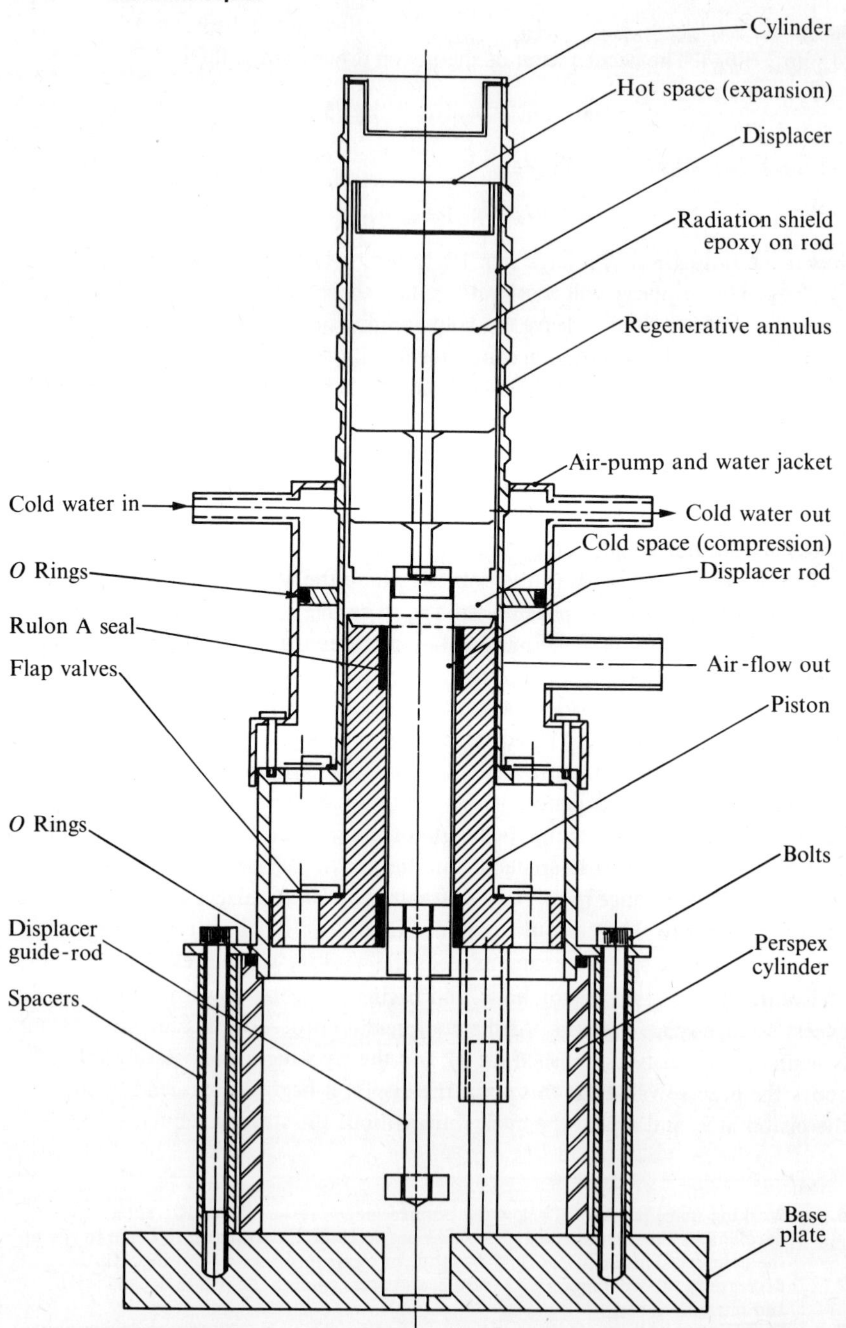

Fig. 10.2. Beale free-piston Stirling engine, arranged as an air compressor.

A conceptual pressure–volume diagram for the system is shown on Fig. 10.1. In practice, steady conditions are not attained in onc cycle, as described here.

Applications of the Beale engines

Work can be extracted from Beale engines by attaching a load to the oscillating piston. Fig. 10.2 shows one form of Beale engine, arranged as a gas compressor. The piston and cylinder of the gas compressor are symmetrically arranged around the engine's piston and cylinder. Agbi (1971) carried out systematic studies of an engine of this type. A characteristic result for the piston and displacer motion, the periodic pressure fluctuation, and the over-all pressure–volume diagram are shown in Fig. 10.3.

In another configuration, shown in Fig. 10.4, the Beale engine can be constructed, so that a light-weight cylinder case and light-weight displacer are

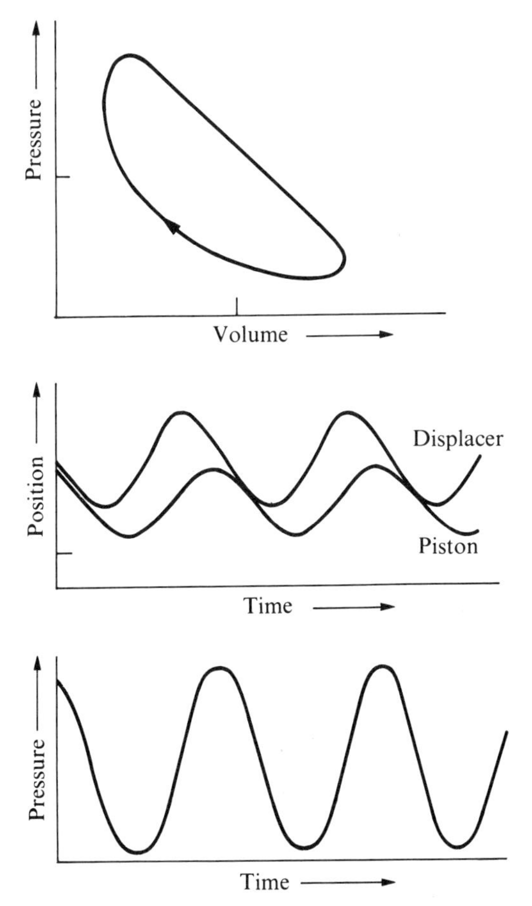

Fig. 10.3. Characteristics of the Beale free-piston Stirling engine air-compressor (After Agbi 1971).

combined with a very heavy piston. The case is mounted in guides to ensure controlled motion. In this cse, the cylinder and the displacer oscillate, while the piston remains stationary. The bottom end of the cylinder can be attached to the ram of an hydraulic pump, and the upper end heated by a combustion flame or solar concentrator. In this form, Professor Beale produced a very effective solar-powered water pump.

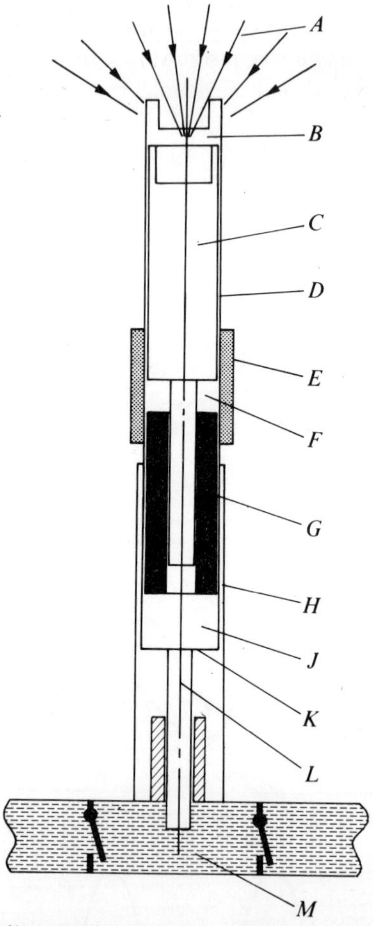

Fig. 10.4. Conceptual diagram of a solar-powered free-piston Stirling engine water-pump. *A* – solar concentrator, *B* – expansion space, *C* – displacer, *D* regenerative annulus, *E* – cooling coils, *F* – compression space, *G* – piston, *H* – cylinder guides, *J* – bounce space, *K* – cylinder, *L* – pump ram, *M* – flap-valve pump.

Other possibilities include a magnet and coil arrangement to extract electric power from the system, or a duplex arrangement where a free-piston engine drives a free-piston cooler, so that a simple tube, heated at one end, becomes cold at the other. Domestic and industrial furnaces, powered by oil or natural gas, frequently require a small electrical supply to drive a fan, or water pump. Serious difficulties arise during an interruption in the electricity supply, since the furnace cannot be used, despite the fact that the gas or oil (contributing 99·9 per cent of the energy) is still available. There is, therefore, a need for a thermally-activated drive system, to replace the electric motor. Normal combustion of fuel would be adequate to energize the unit. An important feature is that the *thermal efficiency does not matter,* because the fuel is burned primarily to provide heating. In this application, initial cost, reliability, and self-starting capability are the important criteria. It would seem to be an ideal application for Beale engines.

Hybrid free-displacer crank-controlled piston engine

Research on Beale free-piston engines stimulated the invention, at the University of Calgary, of a hybrid machine with a crank-controlled piston and a free displacer. Beale free-piston engines are highly attractive from the aspect of absolute simplicity. However, unless they are used as pumps or compressors, it is difficult to utilize the reciprocating motion, because so many engineering operations are geared to rotating shafts.

The hybrid machine is an attempt to combine the advantage of free-piston simplicity with the utility of a piston and crankshaft. One attractive advantage of the hybrid machine is that the bottom end, the piston, cylinder, and crank-connecting-rod assembly can be all standard internal-combustion engine parts. The accumulated production experience, the jigs, tools, and fixtures of conventional internal-combustion engines can be used, permitting a reduction in the cost of Stirling engines to levels nearly comparable with other piston engines.

Fig. 10.5 is a section diagram of a single-cylinder hybrid engine. The crankcase, crankshaft, connecting rod, cylinder and piston could all be conventional gas- or oil-engine components. The piston is modified to include a gas-tight seal, it carries an axial vertical extension which supports a dummy piston, and above that acts as a hollow displacer rod. The space between the piston and dummy piston is connected to the displacer's interior, and, together, these constitute the 'bounce space'. The working space is the volume contained in the cylinder above the dummy piston, and is divided (by the displacer) into the compression space, below the displacer, and the expansion space, above the displacer.

The machine operates in precisely the same way as the Beale free-piston engine, described in the previous section. The light-weight displacer responds rapidly to differences in pressure between the bounce space and the working space. The piston, crankshaft, flywheel and connecting-rod assembly all combine to give the dynamic equivalent of the heavy piston, with high inertia, required by the Beale engine.

Fig. 10.6 shows a research engine of the hybrid variety, built on the crankcase of a small single-cylinder industrial Honda engine. It was constructed at the

University of Calgary. This engine was purchased complete for $97 (in 1970) and at least half the equipment was removed, to strip the engine to a form suitable for conversion. So far, the hybrid engine, with electric-resistance heating, has not operated successfully, because, after a few revolutions, the piston and displacer go out of phase. The engine has no preferred direction of rotation, and operates equally well (or, perhaps, badly), in either direction. Furthermore, at this early stage, it has the disconcerting characteristic of stopping rotation in

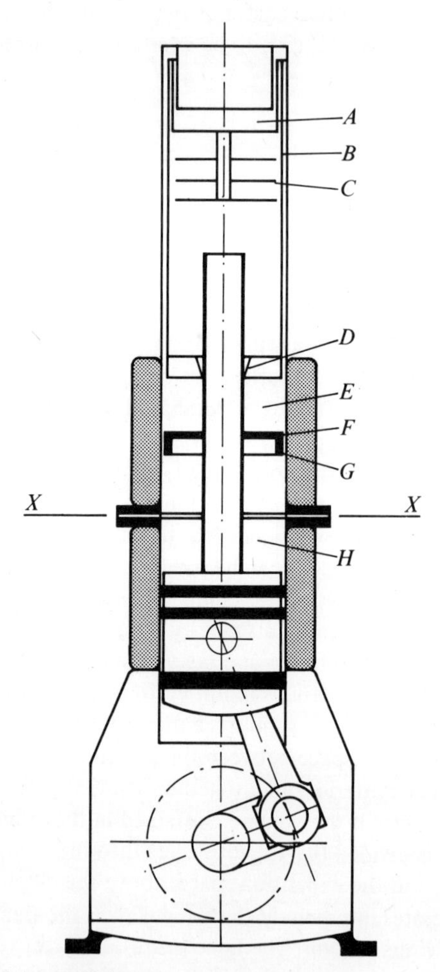

Fig. 10.5. Hybrid crank-controlled piston free-displacer engine.
A – expansion space, *B* – displacer, *C* – radiation shields, *D* – displacer seal, *E* – compression space, *F* – dummy piston, *G* – piston seal, *H* – bounce space.

This machine is dynamically similar to a Beale engine with the piston, connecting-rod, crankshaft, and flywheel all contributing to the effective mass of the piston. It provides for work to be extracted by means of a rotating shaft. A conventional internal-combustion engine may be used below the line *XX*.

one direction and restarting in the other. It is expected that all these problems will be resolved, as further understanding of the engine is gained.

Fig. 10.7 is a view of future possibilities for the hybrid machine. It shows the cross-section of a standard V8 petrol engine, converted to operate as a hybrid. There is a central single-chamber aircraft-type combustor, coupled to each

Fig. 10.6. Prototype hybrid engine, built on Honda industrial engine crankcase (University of Calgary).

cylinder by a liquid-metal heat-pipe. The bounce spaces of all cylinders are interconnected to provide, in effect, a very large gas-reservoir space. The working fluid might be air, pressurized to a few atmospheres, with the engine rated at, maybe, 20 to 30 horsepower, a tenth of its rating as a petrol motor. It has an over-all efficiency of, perhaps, 20 per cent.

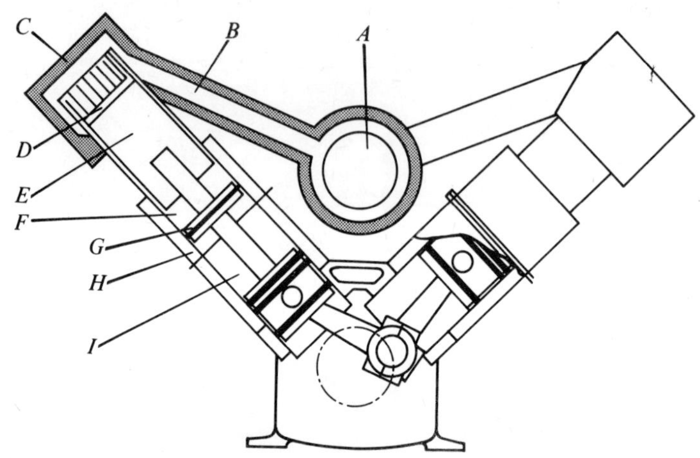

Fig. 10.7. Conceptual diagram of hybrid engine, based on conventional V-engine assemblies. *A* – combustion chamber, *B* – liquid-metal heat-pipe, *C* – thermal insulation, *D* – expansion space, *E* – free displacer, *F* – compression space, *G* – piston, *H* – cooling system, *I* – bounce space.

Compound two-component two-phase working fluids

Two-component two-phase working fluid is being studied at the University of Calgary, with a view to using it for increasing specific output, as an alternative to using extreme pressurization.

From the Schmidt analysis, presented earlier, it is clear that one way to improve the specific output of a given engine is to increase the pressure, either mean or maximum. It has been verified experimentally that the increase in output is virtually linear with increase in pressure, but there is a slight (but progressive) decrease in the rate, owing to limited heat-transfer and fluid-friction effects. A physical understanding of why the power increases as the pressure increases may be gained by reference to Fig. 10.8. This figure is drawn schematically for an engine of such configuration that a volume ratio V_{max}/V_{min} = 2·0 gives a pressure ratio p_{max}/p_{min} = 2·5. To a first approximation, an increase in the mean pressure does not change the *ratio* of the maximum to minimum pressure, but it does change the *range* of the pressure-excursion. Thus, on the diagram shown, an increase in mean pressure from 1·75 units to three times that value (5·25), causes an increase in the range of the pressure-excursion from the initial value of (2·5 – 1) = 1·5 units to (7·5 – 3) = 4·5 units, an increase of three times. While the *ratio* remains the same, the *range* of the pressure-excursion increases, and the P–V diagram is larger, so that more work is produced, and the specific output is increased.

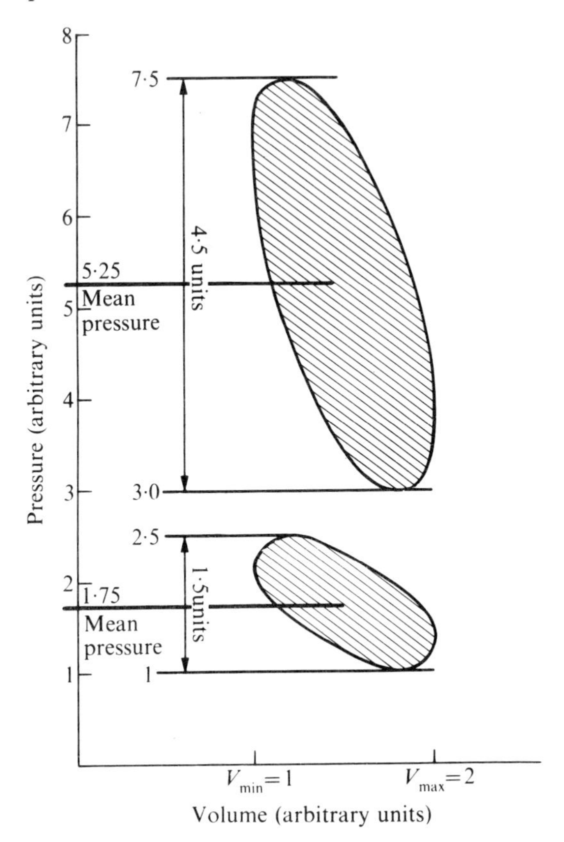

Fig. 10.8. Effect of mean pressure level on range of pressure excursion and engine output. V_{max}/V_{min} = 2:1, p_{max}/p_{min} = 2·5:1. Increase of the *mean* pressure level by a factor of three, from 1·75 to 5·25 units, increases the range of the pressure excursion and the engine output three times.

An advantageous increase in the pressure-excursion could be gained, if it were possible to increase the volume ratio V_{max}/V_{min}. However, it is difficult to design a machine with a ratio much better than 2·5 to 1. With a gaseous working fluid, the only other way to increase the range of the pressure-excursion is to elevate the mean pressure. This is the route followed by Philips, and the result is engines pressurized with hydrogen and helium, to several hundred atmospheres. Another possibility is that an increase in the *range* of the pressure–excursion (but at low mean pressure), might be gained by the use of a working fluid which changes phase in transfer from the compression to expansion space, that is, changes from a liquid to a vapour. For any given fluid at less than the critical temperature, the specific volume of saturated liquid is very much less than for the saturated vapour. Use of a working fluid which experiences a change of phase might produce, therefore, the same effect as an increase in the volume ratio V_{max}/V_{min}, that is, an increase in the range of the pressure-excursion. A speculative impression of the effect is given in Fig. 10.9.

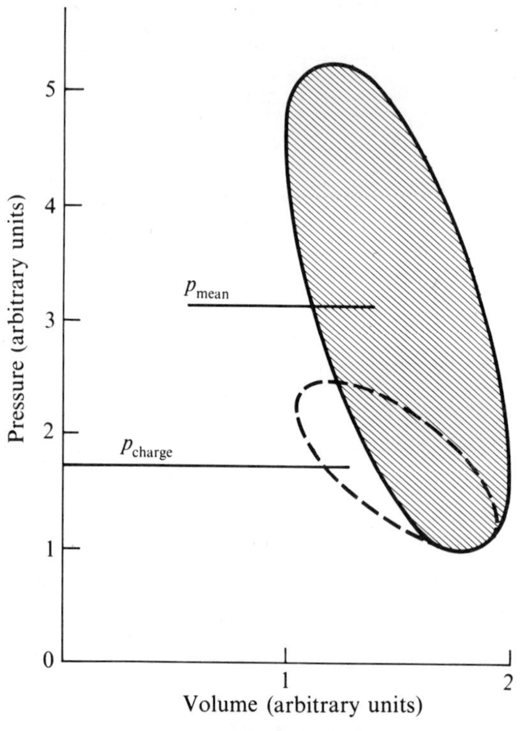

Fig. 10.9. Effect of two-phase two-component working fluid on range of pressure excursion and engine output. A change of phase, from liquid to vapour, of a fraction of the working fluid going from the compression to the expansion space causes an increase in the *effective* volume ratio V_{max}/V_{min}. This causes an increase in the pressure ratio p_{max}/p_{min}. The operating mean pressure p_{mean} is greater than the initial charge pressure p_{charge}, measured with the engine 'cold' and all the phase-change component in the liquid phase. The pressure ratio p_{max}/p_{min} and the range of the pressure excursion are greater than with a single-component single-phase working fluid. The engine output is increased, and there is improved heat transfer during evaporation and condensation.

Supplementary advantages of the two-phase working fluid might include improved heat-transfer and a closer approach to isothermal compression and expansion, resulting from the vapourization and condensation processes.

Preliminary calculations, highly idealized, have indicated that substantial improvement in specific output (of the order of two to three times) might be possible with a compound two-phase two-component working fluid. The fluid at present being studied is water, with air as the gaseous carrier, but there is no reason why a more exotic combination could not be used if it were thermodynamically advantageous and economically worthwhile.

A combination of a hybrid engine (using mainly existing internal-combustion engine parts and operating at moderate mean pressure, with a two phase two component working fluid) might permit the production of an inexpensive high-output Stirling engine.

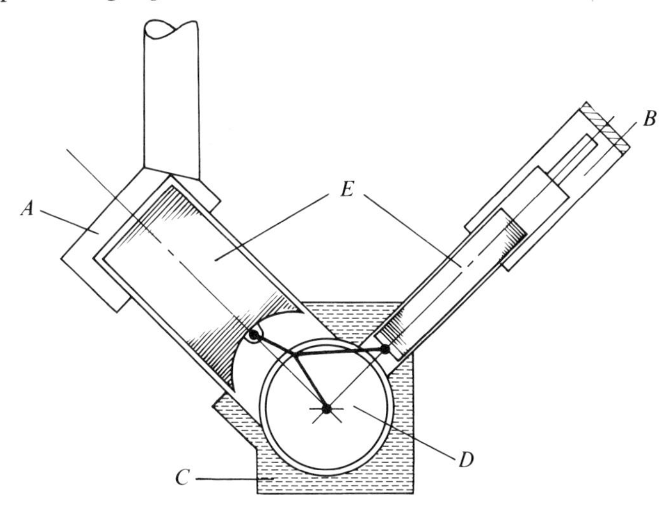

Fig. 10.10. Conceptual diagram for Vuilleumier-cycle machine.
A – heat supplied at high temperature to expansion space.
B – heat subtracted from surroundings at low temperature to expansion space.
C – heat rejected at ambient temperature to cooling system.
D – mechanical work supplied to overcome friction losses.
E – regenerative displacers.

Vuilleumier machine

The name 'Vuilleumier machine' has been applied to a form of thermally-activated cryogenic cooling engine developed recently in the United States, and described by Finkelstein (1970) in reviewing the work of Magee *et al.* (1969) and Pitcher *et al.* (1970).

A Vuilleumier engine is shown diagrammatically in Fig. 10.10, and is similar in concept to the duplex engine described in Chapter 9. It consists of two cylinders, one larger than the other, mounted on a common crankcase. Each cylinder contains a displacer, and the two are coupled by connection to a common crankshaft, so that the displacers move simultaneously but operate out of phase. In the figure, regenerative displacers are shown, but static displacers on one, or both, cylinders could be used as an alternative. External work is supplied to the crankcase to move the displacers. It is important to recognize that the necessary power is very small, since only the fluid-friction and mechanical-friction work is required. External work is *not* required to compress or expand the working fluid, but merely to displace it from one space to another.

In addition to the two regenerators, there are three other heat-exchangers, a heater at the cylinder-head of the larger cylinder, a freezer at the cylinder-head of the smaller cylinder, and a cooler at the lower ends of both cylinders, and also around the crankcase which, in this machine, forms an integral part of the fluid circuit.

The machine operates as follows. As the displacer reciprocates in the large cylinder, the bulk of the working fluid is displaced alternately between the hot, top end of the cylinder and the cold crankcase end. This causes a variation in the pressure of the working fluid; the pressure is a maximum when the fluid is in the hot end, and a minimum when the fluid is in the cold end. The cyclic variation in pressure is utilized in the smaller cylinder, to achieve the desired cooling effect at the top end of the cylinder. The displacer in the smaller cylinder is caused to move, so that it is at the crankcase end of the stroke (i.e. fluid is in the freezing, top end of the cylinder), during 'expansion', when the pressure is changing from its maximum to its minimum value as a result of displacer motion in the large cylinder.

The Vuilleumier machine is difficult to classify. It is really a combination of two Ericsson-cycle machines of the single-cylinder regenerative displacer type, having a constant working volume and an external drive; and yet it has no valves so that, according to the arbitrary definition adopted earlier, it falls into the Stirling-cycle category. It is similar to the duplex Stirling-cycle engine, except that there are no pistons, so that the number of moving parts is halved, a singular advantage. The principal disadvantage is that some external drive is required, albeit of a low power level. Furthermore, the absence of pistons may limit the degree of pressure variation that can be accomplished, and, hence, the cooling capacity. Vuilleumier machines appear less suitable than duplex engines for development as compact high-speed cooling machines, and are not self-contained, but offer many alternative attractions on grounds of simplicity; lack of pistons and seals being the primary advantages.

Regenerator and heat-exchanger design

It is clear (see Chap. 7) that existing design methods for regenerative (and other) heat-exchangers are not satisfactory. Research in this field can be undertaken in university engineering departments, but must be done at a level of understanding sufficiently high for the results to be applicable to the practical problem; the technical literature is littered with studies of regenerative heat-exchangers that have no relevance to Stirling-cycle machines.

Initially, experimental studies of regenerators are to be preferred, rather than purely theoretical studies, and must include investigations of very short blow-times, cyclic variation over a wide range of pressure, and mass flow and density. They must be conducted with emphasis on the effect of porosity, and with due regard to the effect of fluid-friction losses. Experiments with regenerative matrices, carried out at near-atmospheric pressure and temperature and under slowly-fluctuating conditions, have little, or no, relevance to regenerators in Stirling engines. For studies of regenerative heat-exchangers in Stirling engines, there appears to be no better vehicle than a Stirling engine. Thus, *in situ* experiments are recommended in preference to studies of the regenerator conducted in isolation. Attempts by inexperienced students to build a research engine will almost invariably result in an engine so poor in thermal performance that the best results will be achieved by removal of the regenerator. The alternative is to use a highly-developed machine, and the only one commercially available is the Type A Philips cooling engine. Sufficient of these have now been produced that it is likely that a used machine (or one unsuitable for sale, for some reason) could be obtained, at reasonable cost, for experimental purposes.

Typical problems in this field, which currently need investigation, include the preferred length–diameter ratios for the matrix, packing arrangements, matrix material, the relative significance of matrix heat capacity, fluid-friction losses, surface areas for heat transfer, effects of fluid-density variation, and effects of frequency of flow-reversal on matrix performance. The possibility for similar studies exists for the heater and cooler recuperative exchangers. Materials, fins versus slots versus holes, preferred dimensions, length–diameter ratios, and how to design in pulsating flow with wide changes in pressure, density, and mass flow, at relatively high frequency, are also questions on which design guidance is required.

Optimum design charts

Another possible area for university-level research is the production of optimum design charts. Design charts, based on the Schmidt theory, were presented in Chapter 5. The idealizations inherent in the theory are likely to distort the charts, in ways that are not understood, at present. Detailed experimental work is required, therefore, to validate these charts, if possible, and to provide a solid basis for future design.

Seals and bearings

Seals and bearings are important in Stirling engines to a critical degree not equalled in other forms of engine.

Fig. 10.11 shows, diagrammatically, the power characteristic of a Stirling engine, as a function of the working-fluid pressure. The upper line represents the linear relationship of engine power and fluid pressure, predicted by the Schmidt theory. From this must be subtracted ΔP, the loss due to mechanical friction of bearings and seals. If this is assumed to be constant with fluid pressure, the intermediate curve on Fig. 10.11 is obtained. In practice, the loss increases with fluid pressure, and the effects of fluid friction and limited heat-transfer become increasingly significant, so that the lower curve is obtained.

Fig. 10.11 indicates clearly that, for engines having a relatively low fluid pressure, it is vital to minimize bearing and seal friction, to prevent this loss becoming a disproportionate fraction of the engine output. Oil-lubricated bearings are, therefore, indicated, but this compounds the seal problem. It is necessary to keep the working fluid in, to maintain pressure and, thus, engine output. At the same time, if oil is used to lubricate bearings, it is necessary to keep the oil out of the working space, to prevent contamination of the regenerator and heat-exchange surfaces. The effect of contamination is to impose a progressively increasing ΔP on the engine output characteristics, owing to the increased regenerator flow-losses.

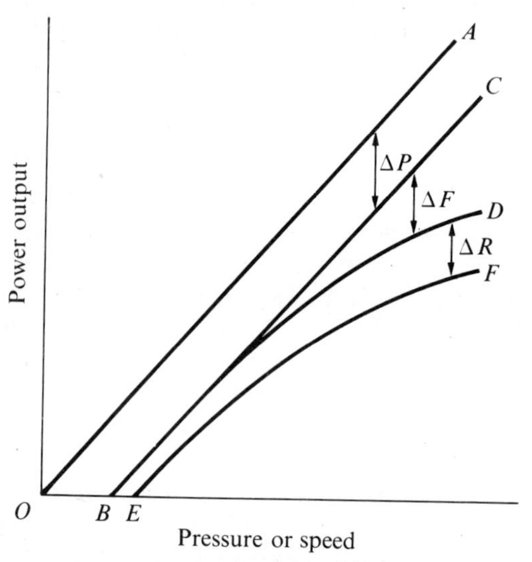

Fig. 10.11. Effect of regenerator blockage on the power characteristic of a Stirling engine. Theoretical power characteristic OA is a straight line from the origin. The mechanical-friction loss ΔP causes the engine-output characteristic to become the line BC. The fluid-friction loss ΔF is a power function of speed, or fluid density, and modifies the engine-output characteristic to BD. Blockage of the regenerator ΔR causes a further decrease in the engine output to EF, and is a progressively cumulative effect.

The question of oil contamination of the working space is a serious one, and to relieve the seal problem it would be better, therefore, to use bearings that are not the conventional oil-lubricated journal bearings. Two possibilities exist: the bearings may be either sealed prelubricated ball (or roller) bearings or unlubricated fluoro-carbon bearings. Sealed prelubricated rolling bearings have relatively-high friction losses, owing to the seals and grease contained within them. They also have a limited life and speed. More important, they cannot be split, so that, for the assembly of a crankshaft big-end bearing, a two-piece composite crankshaft is required. This tends to be expensive, complicated, and weak, relative to a solid crankshaft. Dry lubricated (or unlubricated) fluoro-carbon bearings are more flexible, and allow the use of a variety of arrangements, including split bearings. However, they have comparatively high friction and high rates of wear, compared with conventional oil-lubricated journal bearings. Fluoro-carbon bearing materials are available in a wide range of shapes and sizes, in pure P.T.F.E. or in combination with a variety of filling materials which improve structural integrity. Best results, so far, have been gained with 'RULON', available in England from Crossley (see Appendix).

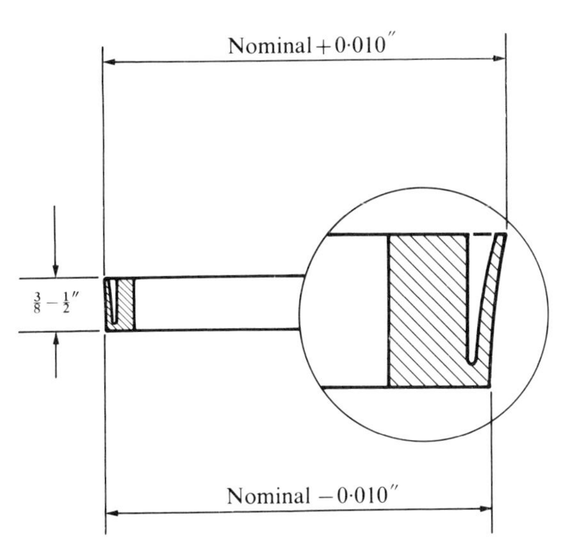

Fig. 10.12. Design details for Rulon seal. (After William Beale, University of Ohio.)

One form of reciprocating gas seal, in Rulon, developed by Professor Beale, is shown in Fig. 10.12. The seal is turned from Rulon bar or sheet stock, and is machined, so that the seal lip is about 0·010 inch oversize on the nominal diameter (with a 2 inch bore). The seal is tapered to a diameter about 0·010 inch undersize nominal, at the lower face. A parting tool is used to cut the seal away, to leave a thin flexible lip about $\frac{3}{8}$ inch long, around the periphery. The seal is mounted on the piston by means of epoxy-cement or a holding plate, and, when inserted into the cylinder, will be very tight. Standing overnight, the seal will become relatively free, and will become loose after working a few times up and down the cylinder. The cylinder should be made of hard metal, honed and polished.

11 Hints for design

The following are some suggestions which merit consideration when designing a Stirling engine.

Be realistic

It is easy to become optimistic about the Stirling engine and enthusiastic about its potential applications. We must recognize that the Philips engine is the result of thirty-five years continuous development by a large team of highly competent engineers, backed by a large international company, in what is, perhaps, the world's leading commercial research establishment. Despite all this, not a single Philips prime mover has been sold on a commercial basis. We must recognize, also, that General Motors, the world's biggest commercial corporation, were licensees of Philips as long ago as 1958, and, after twelve years of considerable effort, abandoned the licence. At the same time, Philips and their two present licensees are commercial organizations obviously stimulated by the profit motive, thus indicating that they must consider Stirling-engine research as worthwhile. Increasing public concern about the environment is resulting in legislation for clean, quiet engines, that will surely increase the price and complexity of internal-combustion engines, and thus provide the opportunity for Stirling engines to compete favourably in automotive applications; in the decade ahead, an interesting situation will develop, the outcome of which is difficult to predict.

Select the target market with discretion

The development of Stirling engines as viable replacements for the petrol and oil engine is a difficult and overcrowded field. Philips are ahead by thirty-five years, and two other groups are also advanced in research. It costs a million pounds plus to become a licensee, thus, this type of development cannot be lightly undertaken.

As an alternative, consider more modest fields, where competition (such as the highly-developed internal-combustion engine) does not exist. There are legitimate applications for Stirling engines in the field of small, unattended, reliable electric-power generators, with chemical, nuclear, and solar power input. For this application, a thermal efficiency of 20 per cent (a reasonably modest value) is adequate, because the only competition is thermoelectric generators at 4 to 7 per cent. Savings in the radioisotope inventory, or the fuel bill, are sufficient to pay for an expensive, hand-built engine. The customers for such a unit are many. They include all the national maritime authorities concerned with navigation aids, civil and military communications departments, and railway companies. The range of interest extends from very small machines (20 to 50 watt electrical output) to machines of several horsepower capacity. The engines should be simple, uncomplicated, and reliable, of modest specific output, and should have a reasonable thermal efficiency.

A market would exist, in undeveloped countries, for a small cheap solar-powered electric generator, capable of charging a battery, to provide light for a few hours at night. Similarly, there seem to be applications for a solar-powered water pump. Stirling engines of up to 10 horsepower (relatively large, quiet, reliable, slow-running, moderately efficient, fossil-fuel burning engines) might prove attractive to yacht owners, for propulsion and generating purposes. Stirling-cycle cooling engines might be used in a variety of refrigerating (or air-conditioning) units for vehicles or buildings, either in the form of electrically-driven or thermally-activated duplex engines.

Avoid complication

The history of Stirling engines contains many examples of ambitious, complicated failures. When entering the field get experience on a small simple single-cylinder engine. Avoid bellows, diaphragms, complicated mechanisms, valves, and materials of unknown quality, all these can be incorporated into subsequent engines if, after some experience, they still appear attractive. One must avoid also the temptation to make the engine design so flexible that virtually any design parameter can be varied over a wide range. This is a common mistake, and, almost invariably, results in a compromised engine, with many flanges and relatively massive thermal-conduction paths.

Heat-exchangers

The rates of heat transfer to and from the engine are critical to the success of the machine. We must separate the hot and cold regions, perhaps using a long regenerative annular duct, as shown in Chapter 7. The thermal conduction paths between the hot and cold spaces must be minimized by the use of long thin wall-sections, and thermal insulation (or radiation and convective baffles) must be provided, to reduce heat losses.

Water cooling should always be used, the heat-transfer coefficient for liquid cooling is orders of magnitude better than for gas-to-wall heat-transfer. Heat transfer between the hot gas and the wall, and between the wall and the working fluid, in both hot and cold regions should be enhanced by the use of extended surface area.

Materials

It is best to use Rulon for seals and bearings. On the original prototypes, do not bother to harden, or plate, the rubbing surfaces on mild steel or stainless steel, but hard-anodize the aluminium rubbing surfaces. Rulon may be lubricated with a smear of light machine oil in preliminary performance trials. It is best to leave the real problem of long-term lubrication until the machine is optimized and into life-evaluation studies.

Avoid the temptation to use hydrogen or helium as the working fluid. Hydrogen is dangerous, and helium is expensive. The use of these gases in anything other than an experimental engine will pose near-insuperable sealing problems. Instead, use air as the working fluid. The sealing problems are still there, but air can be readily replaced by a small engine-driven pump.

Avoid the use of special high-temperature alloy steels for the first prototype. They are expensive, difficult to work, and might prove unnecessary. Stainless steel is usually adequate for the hot region, and mild steel or aluminium for the remainder.

Design thriftily.

Avoid excessive metal-cutting during manufacture, but do not unduly prejudice the design on this account. Frequently, single prototype components are most economically obtained by machining from the solid, whereas, in production, a casting or fabricated component would be used. To reduce the thermal inertia, avoid joints in the high-temperature region; these involve flanges, joints, gaskets, etc., and are best dispensed with by welding simple seams with a low mass of metal involved.

Appendix: Address list

Philips engines

(a) Enquiries to:
  (i) Stirling Engine Product Development Group, or
  (ii) Cryogenerator Division,
    Philips Research Laboratories,
    Eindhoven, Netherlands,
  and in England,
  Mr. D. Handley, Manager,
  Philips Cryogenic Departments,
  Pye Unicam, York Street,
  Cambridge. CB1 2PX.

(b) Reprints of Philips papers may be purchased from
  NV Uitgeversmaatshappij Centrex,
  Nw Emmasingel 9,
  P.O. Box 76,
  Eindhoven, Netherlands.

(c) Philips Licensees
  (i) United Stirling (Sweden) A.B. and Co.,
    Malmo 1, Sweden.
    (Managing Director – Mr. Stig. Carlqvist)
  (ii) Entwicklungsgruppe Stirlingmotor M.A.N./M.W.M.,
    (1) M.A.N. Werk, Augsberg,
  or (2) Motoren-Werke, Mannheim, West Germany.

Small teaching and demonstration Stirling-cycle engines

(a) Radan Associates Limited,
  20 Grove Street, Bath, U.K.

(b) Leybold-Heraeus Limited,
  Blackwall Lane, London, S.E.10.

(c) Cussons Limited,
  102 Gt. Clowes Street, Manchester.

(d) Bradley Engineering,
147 Knoll Drive, Styvechale, Coventry. CV3 5DF.

(e) Drawings for small Heinrici engines may be obtained from

Model Aeronautical Press Limited,
13/35 Bridge Street, Hemel Hempstead, Herts. – Price 50p.

Castings for this engine (price £9·25 per set) may be obtained from

A. J. Reeves and Co.,
416 Moseley Road, Brimingham. B12 9AX.

(f) Small single-cylinder free-piston Stirling engines may be available from

Professor W. Beale,
Department of Mechanical Engineering, University of Ohio,
Athens, Ohio, U.S.A.

Stirling-engine digital simulation programme

Dr. T. Finkelstein,
Trans. Computer Associates, P.O. Box 643, Beverley Hills, California, U.S.A.

Centres where research is known, or is thought to be, in progress on Stirling engines

(a) Professor G. Walker,
Department of Mechanical Engineering, University of Calgary,
Alberta, Canada.

(b) Professor W. Beale,
Department of Mechanical Engineering, University of Ohio,
Athens, Ohio, U.S.A.

(c) Professor J. Smith, Jnr.,
Department of Mechanical Engineering, Massachusetts Institute of Technology,
Cambridge, Mass., U.S.A.

(d) Professor W. Gifford,
Department of Mechanical Engineering, University of Syracuse,
New York, U.S.A.

(e) Professor P. Dunn,
Department of Applied Science, University of Reading,
Reading, Berks, U.K.

(f) Mr. Horace Rainbow,
Teesdale Tools Limited,
192 Overndale Road, Bristol, U.K.

(g) Dr. W. R. Martini, Donald Douglas Laboratories, McDonnel Douglas Company,
Richland, Washington, U.S.A.

Materials

(i) Rulon: in England, Henry Crossley (Packings) Ltd.,
P.O. Box 7, Mill Hill, Astley Bridge,
Bolton. BL1 6PB. Tel. 41351,
in USA, Dixon Corporation,
in Canada, Johnson Plastics.

(ii) Other P.T.F.E. materials:
I.C.I. Limited (Plastics Division),
P.O. Box No. 6, Bessemer Road,
Welwyn Garden City,
Herts. Tel. 23400.
(Tech. Service Note F 13 'Filled P.T.F.E – Properties and Application Design Data'.)

Development

The Intermediate Technology Development Group,
9 King Street, London. WC2E 8HN. Tel. 01 836 5211.
(C. R. Tett (Chief Executive) has published a *Bibliography of Stirling engines.*)

Bibliography

Agarwal, P. D., Mooney, R. J. and Toepel, R. P. (1969). Stirlec 1, a Stirling electric hybrid car. *S.A.E. Paper 690074.*

Agbi, T. (1971). *The Beale free-piston engine.* M.Sc. Thesis, University of Calgary.

Anon. (1968). Smogless Stirling engine promises new versatility. *Prod. Eng.* (Feb.) 31–3.

Anzelius, A. (1926). Über erwarmung vermittels durchstromender medien. *Z. angew. Math. Mech.* **6**, 291–4.

Bahnke, G. D. and Howard, C. P. (1964). The effect of longitudinal heat conduction on periodic-flow heat-exchanger performance. *J. Engng Pwr.* **A86**, 105–20.

Bayley, F. J., Edwards, P. A. and Singh, P. P. (1961). The effect of flow pulsations on heat transfer by forced convection from a flat plate. *Int. Heat. Trans. Conf., A.S.M.E., Boulder, Colorado, U.S.A.* 494–509.

Beale, W. (1969). Free-piston Stirling engines – some model tests and simulations. *S.A.E. Paper 690230, S.A.E. Auto Eng. Congr., Detroit, U.S.A.*

Boestad, G. (1938). Die wärmeubertragung im Ljungstrom luftwärmer. *Feuerungstecknik.* **26**, p. 282.

De Brey, H., Rinia, H. and van Weenan, R. L. (1948). Fundamentals for the development of the Philips air-engine. *Philips. Tech. Rev.* **9**, 97–104.

Buck, K. E. (1968). Experimental efforts in Stirling engine development. *A.S.M.E. Paper No. 68-WA-Ener 3.*

Cayley, G. (1807). *Nicholson's Journal* (November). p. 206 (letter).

Coppage, J. (1952). *Heat-transfer and flow-friction characteristics of porous media,* Thesis, Stanford University, U.S.A.

Coppage, J. E. and London, A. L. (1953). The periodic-flow regenerator -- a summary of design theory. *Trans. Am. Soc. Mech. Engrs.* **75**, 779–87.

Coppage, J. E. and London, A. L. (1956). Heat-transfer and flow-friction characteristics of porous media. *Chem. Engng Prog.* **52**, No. 2 (Feb.), 56–7.

Creswick, F. A. (1957). A digital computer solution of the equation for transient heating of a porous solid, including the effects of longitudinal conduction. *Ind. Math.* **8**, 61–8.

Creswick, F. A. (1965). Thermal design of Stirling-cycle machines. *S.A.E. Paper 949C, Int. Auto Eng. Congr. Detroit, U.S.A.*

Daniels, A. and Du Pre, F. K. (1971). Miniature refrigerators for electronic devices. *Philips Tech. Rev.* **32**, No. 2, 49–56.

Darling, G. B. (1959). Heat transfer to liquids in intermittent flow. *Petroleum,* (May) 177–8.

Davis, S. J. and Singham, J. R. (1951). Experiments on a small thermal regenerator. *General discussion on heat transfer.* I.Mech.E. London, pp. 434–5.

Denham, F. R. (1953). *A study of the reciprocating hot-air engine,* Durham University Ph.D. Thesis.

Dros, A. A. (1965). An industrial gas refrigerating machine with hydraulic piston drive. *Philips Tech. Rev.* **26**, 297–308.

Emerson, D. C. (1959). *Effect of regenerator matrix arrangements on the performance of a gas refrigerating machine.* Durham University B.Sc. Hons. Thesis.

Fabbri, S. (1957). Hot-air engines and the Stirling cycle. *Metano.* **11**, 1–9.

Finkelstein, T. (1952). *Theory of air cycles with special reference to the Stirling cycle.* Ph.D. Thesis, University of London.

Finkelstein, T. (1953). *Self-acting cooling cycles.* D.I.C. Thesis, Imperial College, London.

Finkelstein, T. (1959a). Air engines. *Engineer.* **207**, 492–7, 522–7, 568–71, 720–3.

Finkelstein, T. and Polanski, C. (1959b). Development and testing of a Stirling-cycle machine with characteristics suitable for domestic refrigeration. *English Electric Report W/M(3A).U.5.*

Finkelstein, T. (1961a). Generalized thermodynamic analysis of Stirling engines. *S.A.E. Paper 118B (Annual winter meeting, Detroit).*

Finkelstein, T. (1961b). Optimization of phase angle and volume ratios in Stirling engines. *S.A.E. Paper 118C (Annual winter meeting, Detroit).*

Finkelstein, T. (1961c). Regenerative thermal machines. *Battelle Tech. Rev.* (May).

Finkelstein, T. (1961d). Conversion of solar energy into power. *A.S.M.E. Paper No. 61-WA-297 (Annual winter meeting, New York).*

Finkelstein, T. (1962). Cyclic processes in closed regenerative gas machines analysed by a digital computer, simulating a differential analyzer. *Trans. Am. Soc. Mech. Engr.* **B84,** No. 1 (Feb.).

Finkelstein, T. (1963). Analogue simulation of Stirling engines. *Simulation.* No. 2 (March). (Western Simulation Council Meeting, Los Angeles.)

Finkelstein. T. (1964a). Analysis of practical reversible thermodynamic cycles. *Paper No. 64-HT-37, Jt. A.I.Ch.E. and A.S.M.E. Heat Trans. Conf., Cleveland, Ohio.*

Finkelstein, T. (1964b). Specific performance of Stirling engines. *Third Conference on Performance of High Temperature, Systems, Pasadena, California.*

Finkelstein, T. (1965). Simulation of a regenerative reciprocating machine on an analog computer. *S.A.E. Paper 949F (Annual winter meeting, Detroit).*

Finkelstein. T. (1967a). Thermophysics of regenerative energy conversion. *A.I.A.A., Paper No. 67–216 (5th Aero Science meeting, New York).*

Finkelstein, T. (1967b). Thermodynamic analysis of Stirling engines. *J. Spacecraft & Rockets.* **4**, No. 6.

Finkelstein, T. (1970a). Thermocompressors, Vuilleumier and Solvay machines. *5th I.E.C.E.C., Las Vegas, Nevada, U.S.A. (Sept.).*

Finkelstein, T., Walker, G. and Joschi, J. (1970b). Design optimization of Stirling-cycle cryogenic cooling engines, by digital simulation. *Paper K4, Cryogenic Engineering Conference, Boulder, Colorado, U.S.A. (June).*

Fleming, R. B. (1962). *An application of thermal regenerators to the production of very low temperatures.* Sc.D. Thesis, Massachusetts Institute of Technology, U.S.A.

Flynn, G., Percival, W. H. and Heffner, F. E. (1960). The G.M.R. Stirling thermal engine. *S.A.E. Paper 118A.*

Furnas, C. (1932). Heat transfer from a gas stream to bed of broken solids. *Bull. U.S. Bur. Mines,* No. 361.

Gamson, B. W., Thodos, G. and Hougen, O. A. (1963). Heat, mass and momentum transfer in the flow of gases through granular solids. *Trans. Am. Inst. Chem Engrs.* **39**, 1–35.

Gifford, W. E. and Longsworth, R. C. (1964). Pulse-tube refrigeration progress. *Advances in cryogenic engineering,* Vol. 10, Section M–U. Plenum Press, New York. pp. 69–79.

Gifford, W. E. and Longsworth, R. C. (1965). Surface heat pumping. *Cryogenic Engineering Conference, Rice University, Houston, Texas, U.S.A.*

Glassford, A. P. M. (1962). An oil-free compressor, based on the Stirling cycle. M.Sc. Thesis, Dept. of Mech. Eng., M.I.T., U.S.A.

Goranson, R. B. (1968). Development of a simplified Stirling engine to power circulatory-assist devices. *Proceedings of the 3rd Intersociety Energy Conversion Conference, Boulder, Colorado, U.S.A. (Aug.).*

Grashof, F. (1890). *Theorie der kraftmaschinen.* Hamburg.

Hahnemann, H. (1948). Approximate calculation of thermal ratios in heat-exchangers including heat conduction in the direction of flow. *National Gas Turbine Establishment Memorandum 36.*

Halley, J. A. (1958). The Robinson-type air engine. *J. Stephenson Engng Soc. King's Coll. Newcastle.* **2**, No. 2, p. 49.

Harmison, L. T., Martini, W. R., Rudnick, M. I. and Huffman, F. N. (1972). Experience with implanted radioisotope-fuelled artificial hearts. *EN/IB 10, Proceedings of Second International Symposium on Power from Radio-isotopes, O.E.C.D., Madrid (June).*

Harper, D. B. and Rohsenow, W. M. (1953). Effect of rotary regenerator performance on gas-turbine-plant performance. *Trans. Am. Soc. mech. Engrs.* **75**, 759–65.

Hausen, H. (1929). Über die theorie des warmeaustausches in regeneratoren. *Z. angew Math. Mech.* **9** (June), 173–200. (On the theory of heat exchange in regenerators. *R.A.E. Library Translation, No. 126.*)

Hausen, H. (1931). Naherungsverfahren zur berechnung des warmeaustausches in regeneratoren. *Z. Angew Math. Mech.* **11** (April), 105–14. (An approximate method of dimensioning regenerative heat-exchangers. *R.A.E. Library Translation, No. 98*).

Hausen, H. (1942). Vervollstandigte berechnung des warmeaustauches in regeneratoren. *Z. ver. dt. ing. beiheft Verfahrenstechnik.* No. 2, p. 31. (*M.A.P. Reports and Translations, No. 312 (1946).*)

Havemann, H. A. and Narayan Rao, N. N. (1955). Studies for new hot-air engine. *J. Indian Inst. Sci.* **B37**, p. 224, and **38**, p. 172.

Havemann, H. A. and Narayan Rao, N. (1954). Heat transfer in pulsating flow. *Nature.* **174**, No. 4418, p. 41.

Heffner, F. E. (1965). Highlights from 6500 hours of Stirling-cycle operation. *S.A.E. Paper No. 949D, Int. Auto Engr. Congress, Detroit, U.S.A.*

Henderson, R. E. and Dresser, D. L. (1960). Solar concentration associated with the Stirling engine. *A.R.S. Space power-systems Conference (Sept.).*

Herschel, J. (1850). Making ice. *The Athenaeum* (Jan. 5), p. 22.

Hogan, W. H. and Stuart, R. W. (1963). Design considerations for cryogenic refrigerators. *A.S.M.E. Paper No. 63-WA-292.*

Hougen, J. O. and Piret, E. L. (1951). Effective thermal conductivity of granular solids through which gases are flowing. *Chem. Engng Prog.* **47**, 295–303.

Howard, C. P. (1963). Heat-transfer and flow-friction characteristics of skewed-passage and glass-ceramic heat-transfer surfaces. *A.S.M.E. Paper No. 63-WA-115.*

Howard, C. P. (1964). The single-blow problem including the effects of longitudinal conduction. *A.S.M.E. Paper No. 64-GTP-11.*

Hurley, E. G. (1954). Tests on a twin piston Stirling-cycle engine, using internal combustion. *Shell Thornton Report K. 121.*

Iliffe, C. E. (1948). Thermal analysis of the contra-flow regenerative heat-exchanger. *Proc. Instn mech. Engrs.* **159**, 363–72.

Jakob, M. (1957). *Heat transfer,* Vol. II. John Wiley and Sons, New York.

Johnson, J. E. (1952). Regenerator heat-exchangers for gas turbines. *Aerc Research Council Technical Report, R & M No. 2630.*

Jones, L. L. Jnr. and Fax, D. H. (1954). Perturbation solutions for the periodic-flow thermal regenerator. *A.S.M.E. Paper No. 54-A-130.*

Joule, J. (1852). On the air engine. *Phil. Trans. R. Soc.* **142**.

Karavansky, I. I. and Meltser, L. Z. (1958). Thermodynamic investigations of the working cycle of the Philips machine. *Proc. 10th Int. Cong. Refrigeration.* 3--29, 209.

Kays, W. M. and London, A. L. (1958). *Compact heat-exchangers.* McGraw-Hill, New York.

Kirk, A. (1874). On the mechanical production of cold. *Proc. Inst. mech. Engrs.* **37**, 244–315.

Kirkley, D. W. (1959). *Continued work on the hot-air engine.* Durham University, B.Sc. Hons. Thesis.

Kirkley, D. W. (1962a). *An investigation of the losses occurring in reciprocating hot-air engines.* Durham University, Ph.D. Thesis.

Kirkley, D. W. (1962b). Determination of the optimum configuration for a Stirling engine. *J. Mech. Engng Sci.* **4**, No. 3, 204–12.

Kirkley, D. W. (1965). A thermodynamic analysis of the Stirling cycle and a comparison with experiment. *S.A.E. Paper 949B, Int. Auto Engng Congress, Detroit, U.S.A.*

Köhler, J. W. L. and Jonkers, C. O. (1955a). Fundamentals of the gas refrigerating machine. *Philips Tech. Rev.* **16**, 69–78.

Köhler J. W. L. and Jonkers, C. O. (1955b). Construction of a gas refrigerating machine. *Philips Tech. Rev.* **16**, 105–15.

Köhler, J. W. L. (1960). *Prog. in cryogen.* **2**, 41–67.

Köhler, J. W. L. (1965). The Stirling refrigeration cycle. *Scien. Am.* **212**, No. 4, 119–27.

Kolin, I. (1968). The Stirling cycle with nuclear fuel. *Nucl. Eng.* (Dec.), 1027–34.

van der Laan, G. M. J. and Roozendaal, K. (1961). A snow separator for liquid-air installations. *Philips Tech. Rev.* **23**, No. 2, 48–54.

Lambertson, T. J. (1958). Performance factors of a periodic-flow heat-exchanger. *Trans. Am. Soc. mech. Engrs.* **80**, 586–92.

Lienesch, J. H. and Wade, W. R. (1969). Stirling engine operating quietly with almost no smoke and odour, and with little exhaust emission. *S.A.E. Journal.* 40–44.

Locke, G. L. (1950). Heat-transfer and flow-friction characteristics of porous solids. *T.R. No. 10, Dept. of mech. Eng. Stanford University, U.S.A.*

Lucek, R., Damsz, G. and Daniels, A. (1967). Adaptation of rolling-type seal diaphragms to miniature Stirling-cycle refrigerators. *Air Force Flight Development Laboratory, TR.-67-96 (July).*

Magee, P. R. and Datring, R. (1969). Vuilleumier-cycle cryogenic refrigerator development. *Technical Report, TR 68-69 U.S. Air Force Flight Dynamics Lab.*

Malone, J. F. J. (1931). A new prime mover. *Jl. R. Soc. Arts,* Vol. LXXIX. No. 4099, 680.

Martinelli, R. C., Boelter, L. M. K., Winberge, E. B. and Yakahi, S. (1943). Heat transfer to a fluid flowing periodically at low frequencies in a vertical tube. *Trans Am. Soc. mech. Engrs.* **65**, 789–98.

Martini, W. R. (1968). A Stirling-engine module to power circulatory-assist devices. *A.S.M.E. Paper No. 68-WA-Ener. 2.*

Martini, W. R., Johnson, R. P., and Noble, J. E. (1969). Mechanical engineering problems in energetics–Stirling engines. *A.S.M.E. Paper No. 69-WA-Ener 15.*

McMahon, H. D. and Gifford, W. E. (1960). A new low-temperature gas expansion cycle, Parts I and II. *Advances in cryogenic engineering,* Vol. 5. Plenum Press, New York. pp. 354–72.

Meek, R. M. G. (1961). The measurement of heat-transfer coefficients in packed beds by the cyclic method. *Int. Heat-Trans. Conf. (A.S.M.E.), Boulder, Colorado, U.S.A.* pp. 770–80.

Meijer, R. J. (1959). The Philips hot-gas engine with rhombic drive mechanism. *Philips Tech. Rev.* **20**, No. 9, 245–76.

Meijer, R. J. (1960). *The Philips Stirling thermal engine.* Ph.D. Thesis, Technical University, Delft. (Also published as *Philips Research Reports, Supplements,* No. 1 (1961).)

Meijer, R. J. (1965). Philips Stirling engine activities. *S.A.E. Paper No. 949E (Annual winter meeting, Detroit, U.S.A.).*

Meijer, R. J. (1969a). The Philips Stirling engine. *Ingenieur.* **81**, W69–W79, W81–W93.

Meijer, R. J. (1969b). Rebirth of the Stirling engine. *Sci. J.* **A5**, No. 2, 31–9.

Meijer, R. J. (1970). Prospects of the Stirling engine for vehicular propulsion. *Philips Tech. Rev.* **31**, No. 5/6, 169.

Mondt, J. R. (1964). Vehicular gas-turbine periodic-flow heat-exchanger solid and fluid temperature distributions. *J. Engng Pr.* A**86**, 121–6.

Murray, J. A., Martin, B. W., Bayley, F. J. and Rapley, C. W. (1961). Performance of thermal regenerators under sinusoidal flow conditions. *Int. Heat-Trans. Conf., A.S.M.E., Boulder, Colorado and London, England.* pp. 781–96.

Narayan Rao, N. N. (1954). Problems relating to the development of internal combustion engine industry in India, in *A new hot-air engine.* C.S.I.R. (New Delhi) Report, pp. 49–56.

Neelen, G. T. M., Ortegren, L. G. H., Kuhlmann, P. and Zacharias, F. (1971). Stirling engines in traction applications. *C.I.M.A.C., A26, 9th Int. Congress on combustion engines, Stockholm, Sweden.*

Nusselt, W. (1927). Die theorie des winderhitzers. *Z. Ver. dt. Ing.* **71**, 85.

Nusselt, W. (1928). Der beharrungszustand im winderhitzer. *Z. Ver. Dt. Ing.* **72**, 1052.

Otten, E. H. (1956). Tests on a displacer-type Stirling engine using internal combustion. *Shell Thornton Report K. 140.*

Parker, M. D. and Smith, C. L. (1960). Stirling engine development for space power. *A.R.S. Space power-systems Conf. (Sept.).*

Paste, E. A. and Whitaker, R. O. (1961). Investigation of a 3-kW Stirling-cycle solar power system. *WADD-TR-61-122 (in 10 vols) Part 1, Engine design, Part X, Experimental evaluation.*

Pitcher, G. K. and du Pré, F. K. (1970). Miniature Vuilleumier-cycle refrigerator. *Proc. Cryogenic Engineering Conference, Boulder, Colorado, U.S.A.*

Prast, G. (1963). A Philips gas refrigerating machine for 20K. *Cryogenics.* (September) 156–60.

Qvale, E. B. and Smith, J. L. Jnr. (1968). A mathematical model for steady operation of Stirling-type engines. *J. Engng Pwr.* A, No. 1, 45–50.

Qvale, E. B. and Smith, J. L. Jnr. (1969). An approximate solution for the thermal performance of a Stirling engine regenerator. *J. Engng Pwr.* A, No. 2, 109–12.

Rankine, M. (1854). On the means of realizing the advantages of air engines. *Proc. Br. Ass.*

Rapley, C. (1960). *Heat transfer in thermal regenerators.* M.Sc. Thesis, Durham University.

Rietdijk, J. A., Van Beukering, H. C. J., van der Aa, H. H. M. and Meijer, R. J. (1965). A positive rod or piston seal for large pressure differences. *Philips Tech. Rev.* **26**, 287–96.

Rios, P. A. and Smith, J. L. Jnr. (1969). An Analytical and experimental Evaluation of the pressure-drop losses in the Stirling cycle. *A.S.M.E. Paper No. 69-W 69–WA/Ener. 8.*

Romie, F. E. and Ambrosio, A. (1966). Heat transfer to fluids flowing with velocity pulsations in a pipe. *Heat transfer, thermodynamics and education.* McGraw-Hill, New York. pp. 273--94.

Saunders, O. and Ford, H. (1940). Heat transfer in the flow of gas through a bed of solid particles. *J. Iron Steel Inst.* No. 1, p. 291.

Saunders, O. A. and Smoleniec, S. (1948). Heat regenerators. *Proc. 7th Int. Congress Appl. Mech. Vol. 3,* pp. 91–105.

Schalkwijk, W E (1959). A simplified regenerator theory. *J. Engng Pwr.* **A81**, 142--50.

Schmidt, G. (1861). Theorie der geschlossenen calorischen maschine von Laubroy und Schwartzkopff in Berlin. *Z. Ver. Oster. Ing. p. 79.* (1871). Theorie der Lehmann schen calorischen maschine. *Z. Ver. dt. Ing.* **15**, No. 1.

Schultz B. H. (1951). Regenerators with longitudinal heat conduction. *General discussion on heat transfer (I.Mech.E. and A.S.M.E.).*

Schultz. B. H. (1953). Approximate formulae in the theory of thermal regenerators. *Appl. scient. Res. A.* **3**, 165--73.

Schumann. T. E. W. and Voss, V. (1934). Heat flow through granulated material. *Fuel.* **13**, 249–56.

Schumann. T. E. W (1929). Heat transfer to a liquid flowing through a porous prism *J. Franklin Inst.* **208**, 405–16.

Shuttleworth, P (1958). *An experimental investigation of a Stirling-cycle engine.* Durham University, M.Sc. Thesis.

Siegel, R. and Perlmutter, M. (1961). Two-dimensional pulsating laminar flow in a duct with a constant wall temperature. *Int. Heat-trans. Conf. (A.S.M.E.), Boulder, Colorado, U.S.A.* pp. 517–35.

Ster, J. Van der (1960). *The production of liquid nitrogen from atmospheric air using a gas refrigerating machine.* Delft Technische Hochschule Thesis.

Stirling R. (1817). *Improvements for diminishing the consumption of fuel and in particular, an engine capable of being applied to the moving of machinery on a principle entirely new.* British Patent No. 4081.

Tipler W (1947). A simple theory of the heat regenerator. *Tech. Report No. ICT/14, Shell Petroleum Co. Ltd.*

Tipler, W (1948). An electrical analogue to the heat regenerator. *Proc. Int. Cong. of Appl. Mech.* Vol 3, pp. 196--210.

Trayser, D. A. and Eibling J. A. (1966). A 50-watt portable generator employing a solar-powered Stirling engine. *Proc. I.E.C.E.C. Conf.* pp. 1008--16.

Van Nederveen H. B. (1966). The nuclear Stirling engine. *Paper 35, Ind. App. of Isotopic power generators, Joint U.K.A.E.A.–E.N.E.A. Intl. Symp. A.E.R.E., Harwell (Sept.).*

Van Weenan F. L. (1948). Construction of the Philips air engine. *Philips Tech. Rev.* **9**, 125–34.

Vasishta, V (1969). *Heat-transfer and flow-friction characteristics of compact matrix surfaces for Stirling-cycle regenerators.* M.Sc. Thesis, University of Calgary.

Wadsworth J. (1961). An experimental investigation of the local packing and heat-transfer processes in packed beds of homogeneous spheres. *Int. Heat-Trans. Conf. (A.S.M.E.), Boulder, Colorado, U.S.A.* pp. 760–9.

Walker, G. (1961a). The operational cycle of the Stirling engine with particular reference to the function of the regenerator. *J. Mech. Engng Sci.* **3**, No. 4.

Walker G. (1961b). *Some aspects of the design of reversed Stirling-cycle machines.* Ph.D. Thesis, University of Durham

Walker G. (1962). An optimization of the principal design parameters of Stirling-cycle machines. *J. Mech. Engng Sci.* **4**, No. 3.

Walker, G. (1963a). Regeneration in Stirling engines. *Engineer, Lond.* **216**, No. 5631.

Walker, G. (1963b). Density and frequency effects on the pressure drop across the regenerator of a Stirling-cycle machine. *Engineer, Lond.* **216**, 1063.

Walker, G. (1963c). Machining internal fins in components for heat-exchangers. *Machinery, Lond.* **101**, No. 2590.

Walker G. and Khan, M. (1965a). The theoretical performance of Stirling-cycle machines. *S.A.E. Paper 949A (Annual winter meeting, Detroit, U.S.A.).*

Walker, G. (1965b). Some aspects of the design of reversed Stirling-cycle machines. *A.S.H.R.A.E., Paper No. 231 (Annual summer meeting, Portland, U.S.A.).*

Walker, G. (1965c). Regenerative thermal machines – a status survey. *Proc. Am. Power Conf. Vol. XXVII, Chicago.* p. 530.

Walker, G. (1967) Stirling-cycle engines for total-energy systems. *Inst. Gas Tech. Report, Chicago.*

Walker, G. (1969). Dynamical aspects of the rhombic drive for small cooling engines *Advances in Cryogenic Engineering,* Vol. 14 (*Ed. K. Timmenhaus*). Plenum Press, New York.

Walker G. (1968). Military applications of Stirling-cycle machines. *I.E.C.E.C., Boulder, Colorado.*

Walker, G. and Vasishta V. (1971). Heat-transfer and friction characteristics of dense-mesh wire-screen Stirling-cycle regenerators. *Advances in Cryogenic Engineering,* Vol. 16 (*Ed. K. Timmerhaus*). Plenum Press, New York.

Walker, G. and Wan W. K. (1972a). Heat-transfer and fluid-friction characteristics of dense-mesh wire screen at cryogenic temperatures. *Proc. 4th Int. Cryogenic Engineering Conference, Eindhoven, Netherlands (May).*

Walker G. (1972b). Stirling engines for isotope power systems. *Proc. 2nd Intl. Conf. on Power from Radioisotopes, Madrid (June).*

Walker. G. (1972c). Stirling engines – the second coming. *Chart. Mech. Engr.* **19** No. 4, 54–7.

Wan, W. K. (1971). *The heat-transfer and friction-flow characteristics of dense-mesh wire-screen regenerator matrices.* M.Sc. Thesis, University of Calgary.

West, F. B. and Taylor, A. T. (1952). The effect of pulsations on heat transfer – turbulent flow of water inside tubes. *Chem. Engng Prog.* **48**, No. 1, 39–43.

Williamson, J M. (1959). The effectiveness of the periodic-flow heat-exchanger. *English Electric Report, No. W/M(4B).*

Yagi, S., Kunii D., and Wakao, N. (1961). Radially effective thermal conductivities in packed beds *Int. Heat-Trans. Conf. (A.S.M.E.) Boulder, Colorado, U.S.A.* pp 742–9.

Yendall E. F. (1958). A novel refrigerating machine. *Advances in cryogenic engineering,* Vol. 2. Plenum Press, New York. pp. 188–96.

Zeuner, G. (1887). *Technische thermodynamik.* Vol. 1. Leipzig. pp. 347–57.

Index